A BRIEF HISTORY

OF

BUTTE, MONTANA

THE WORLD'S
GREATEST MINING CAMP

INCLUDING A STORY OF
THE EXTRACTION AND
TREATMENT OF ORES
FROM ITS GIGANTIC
COPPER PROPERTIES

Illustrated

By HARRY C. FREEMAN
BUTTE, MONTANA

CHICAGO
THE HENRY O. SHEPARD COMPANY
Printers of The Inland Printer
1900

Standard Book Number 87004-127-4

Lithographed and bound in the United States of America by
ARTCRAFT PRINTERS
1207 E. Front
Butte, Montana 59701
(406) 723-4200

ENGRAVINGS BY
ILLINOIS ENGRAVING COMPANY
CHICAGO

PRESS OF
THE HENRY O. SHEPARD CO.
CHICAGO

INTRODUCTION.

❧ ❧ ❧

OUT of the boundless West from time to time comes literature of every description concerning its resources, development, life, grandeur of scenery and every phase that can possibly serve as a vehicle to relieve the mind. One effort emanates from the pen of the student of events, who sees the unfolding of mighty things which shall leave their imprint upon the future of a great and growing nation. Another purports to be the work of the critic, who, after a superficial study of prevailing conditions, finds much delight in exaggerating the primitiveness of its institutions, the roughness of its life and the depravity of its public morality, with little or no thought as to the obstacles which have been overcome, the rapidity with which events have followed one another. nor the influences which have been thrown about them. Still another is of a commercial character, inspired by the demand for sensational nonsense upon the part of the great newspapers of the East, who still find profit in stigmatizing this new country as abnormally " wild and woolly " in contra-distinction to the " civilized and effete East." The " cattle king," the " copper king," the " silver king " and a dozen other titles are still forced upon the credulity of the uninformed to assist in throwing an air of mystery and awe about this bountifully endowed country and to strengthen the stories of fabulous wealth popularly supposed to be found beneath every rock and along every stream thereof.

Some writers studiously adhere to the path of truthfulness; others assume that truthfulness is the last element required. The result can be but one. The average mind is confused and clouded. The resources of the country are admitted, but the state of society is too unattractive. Large opportunities are conceded, but it means a divorcement from all civilizing influences to grasp them. The delightful healthfulness of its climate is recognized, but the weather is too rigorous. Educational institutions are crude, plodding, and partake nothing of the higher standards. Religious life is stunted and warped, and a thoughtful pulpit and a comfortable church home are impossible of attainment. A thousand things are lacking which are necessary and another thousand are present which must be eliminated to make the country tenable. And thus doth the imagination today perform the functions that should rest with certain knowledge, as much as was the case forty or forty-five years ago, when stories of Western exploration and discovery were beginning to work themselves from thence. At that time but little was known of true conditions. From California had come stories of great wealth and, in due course of time, the bones of many a hardy adventurer lay bleaching along the overland trails to guide other courageous spirits toward the setting sun. Fremont's expedition had added a little cumulative testimony to that of daring explorers who had previously sought the source of the great Missouri, but which still left to the imagination the task of adding all the details in arriving at any given fact concerning the whole West. The Mormons had shut themselves in along the banks of the Jordan and about the shores of Great Salt Lake, and details of their fanatical crimes ofttimes carried with them meager facts concerning the country contiguous, but to imagination was left the duty of setting the frame. As it was then, so it is now to almost as great an extent.

Misinformation has erected an average opinion concerning the Great West quite as much at variance with the true conditions as lack of information in the past has done. The West has developed so rapidly and transition from condition to condition has so speedily followed one another that today a new West is presented while the world is still wrestling with the traditions and the legends of the old. While the East is straining its eyes to catch a glimpse of some evidence of a higher degree of civilization, the unsatisfied traveler is wearing himself out in a vain search for lingering relics of primitive life.

Yet all seems to be the part of God's economy,

and logic approves of the enveloping of true conditions in a certain mystery, which shall be dispelled by slow stages of discovery and development in the working out for the whole nation of a destiny palpably intended for it. It furnishes not only a school to the brawn and brain of coming generations, as in the past, but, equally as necessary, perhaps, it supplies a reserve of treasure which shall be at the disposal of the whole nation when most needed.

Step by step have the borders of civilization been pushed from the banks of the Mississippi and the shores of the Pacific until they have merged into one. Gradually have the agricultural, grazing and mineral resources of the Western plains, valleys and mountains been developed until today they are the great producers of raw materials for the gigantic industries of the East. And, so surely, in due time will the industries of the East come creeping westward to utilize these materials at the point of production, while in their wake will come the people of a congesting East. But it will all come in God's time. It will come when an increasing national vigor is vitally necessary. When the voice of power of a great nation in the events of the world must needs be reinforced by the best manhood, by the highest industrial attainment, by the greatest material wealth and by the broadest civilization. How better could this end be reached than by the methods which at present obtain? What better school for the development of the sturdiest, the best that manhood should know for the strenuous struggle for supremacy of a whole nation than the trials and hardships consequent upon the settlement of a great expanse like our West? It was the same school where was learned the spirit of the Revolution which gave us the Republic, and which perpetuated the Republic in the Civil War, and it is the same school which will develop the youth of coming generations who shall stand as sponsors for the Republic's integrity for all time.

There they will go on, seeking out the dangers and the hardships, redeeming the dark, forbidding places, developing and expanding the resources of the country until the East shall know no line of distinction, can see no flaw in its institutions and its civilization, and the best in customs and morality of the one shall be engrafted into the lives and the minds of the people of the other; when the East shall be more Western and the West more Eastern. It is a consummation much to be desired, a condition some day certain of realization. It is the most pregnant promise that is presented to its people at the dawn of the new century of a continued survival and growth of the Republic unto the time when its voice shall be the most potent and its influence the most far reaching of all the nations of the earth. All honor to the West from whence beckoneth the star of empire to the youth of the East and the whole world — not to an empire where royalty reigns, but to a free country where brain and brawn are kings and where determination to do is a more priceless treasure than much fine gold.

That which follows is a story touching upon one of the great landmarks of the West. Here and there others have been erected which, in slight measure only, point what the future has in store. Many States of the great West enjoy such landmarks. They indicate the slow, certain development of the great industries of that great expanse. Still other States are but awakening to a realization of latent possibilities. A generation or so hence thousands, aye millions, of acres of arid lands, rendered, it once seemed, useless, will be reclaimed and put to the plow by the agency of irrigation, and Kansas and the Dakotas will be met by lusty rivals in new grain-bearing States. Stretching along the great Rockies from border to border discoveries are fast being made which tend to identify the whole range as a vast storehouse of mineral wealth. Great camps have sprung into existence whose futures for long years to come are assured. Some are gold camps, others silver, but that of which our story deals is a copper camp.

If the same elements had controlled the development of Butte as have shaped the destinies of other equally promising mining-camps, its end would, no doubt, have been as inglorious. Denuded, as it seemed, of all the wealth that nature had hidden beneath its surface and rendered unattractive as a source of further treasure, it seems nothing short of marvelous that the camp was not abandoned for at least a long cycle of years — perchance forever — unmarked save by the tell-tale ruins of its early exploitation.

Situated in an almost inaccessible valley, shut in by an abrupt curve of the Rocky Mountains and off-running spurs and foothills, it most certainly would have been least sought in the pursuit of all the engagements of the human race but for that one industry which has made its fame world-wide as the greatest city of its kind on earth, namely, mining. Mineral wealth was there and in abundance. God seems even to have allowed the scale of equal distribution to go sadly out of horizontal in his endowment of that small area of hills which surround Butte proper, from which have been taken the riches of an empire and which are yet but in the babyhood of their development. But upon their discovery hinges the most remarkable feature of the story of Butte, aside from the unequaled story of its wonderful development and growth and its present wealth and pregnant future.

It is with regret that the following contents are, of necessity, confined to the one city of Butte. So great are the other resources of the whole State of Montana that a recital of them all would immeasurably add to the value of the work in dispelling erroneous ideas concerning the commonwealth in particular and the whole West in general and create a more healthful opinion of the same in the minds of the uninformed. The great sheep-raising industries of the State surpass overwhelmingly those of any other State in the Union; consequently this is true of wool. On a thousand ranges are fattened the cattle whose delicious qualities the whole world knows, and herein is presented an industry closely rivaling any other State, and so advantageously endowed is the State in this respect that a matter of a few years will place it at the head in this industry. No richer agricultural lands can be found the country over than along many of the valleys of the State, and especially is this true of the Gallatin and Bitter Root valleys, whose fame has crossed the borders of the State, which present opportunities of the greatest magnitude. Irrigation is rapidly reclaiming large portions of the State for agricultural purposes and, when the fact is realized that the products of the State from this source are wholly inadequate to supply the needs of home consumption, the advantages here presented are palpable. Mining is being largely developed along the whole length of the Rockies and off-running ranges throughout the State and opportunities in this direction have but had their surface pricked.

In compiling the matter for this work the idea has been to create a healthful opinion and erect a curiosity for a deeper knowledge of the subject treated and present to the people of the city and State something that will adequately do justice to one phase of Montana's resources and prospects. To accomplish this it has been considered wise to depart from too dry details and wearisome statistics, seeking to encourage the reader to peruse its entire contents so that, at its conclusion, he may be forced to the admission that something new has been revealed and a desire excited for further facts concerning the great West. To this end the following humble effort is respectfully submitted.

HARRY C. FREEMAN.

BUTTE, MONTANA, November 17, 1900.

TABLE OF OUTPUTS OF LEADING MINING STATES.

No.	State.	Coal.	Iron.	Copper.	Gold.	Silver.	Total Output, including Lead and Zinc.
1	Michigan................	$ 720,000	$186,433,371	$27,444,442	....	....	$214,597,813
2	Pennsylvania............	161,209,231	....	....	....	....	161,209,231
3	**Montana**................	**2,227,998**	....	**40,941,906**	**$4,819,157**	**$21,786,834**	**68,447,309**
..	**Butte's product**........	....	....	**40,882,492**	**1,282,447**	**12,742,893**	**54,907,853**
4	Colorado................	8,471,105	....	1,869,169	26,508,675	13,771,731	56,791,425
5	Arizona.................	....	....	22,079,023	2,575,000	1,191,600	25,845,623
6	Illinois................	18,443,946	....	....	....	....	20,343,682
7	California..............	430,631	....	4,211,517	14,800,000	357,480	19,799,628
8	Ohio....................	14,191,557	....	....	....	....	14,191,557
9	West Virginia...........	11,830,773	....	....	....	....	11,830,773
10	Utah....................	....	....	1,639,550	3,506,582	4,279,695	9,425,827
11	Idaho...................	....	....	....	1,750,000	2,859,840	9,180,376
12	Alabama.................	7,484,763	....	....	....	....	7,484,763
13	Kansas..................	5,124,248	....	....	....	....	7,277,554
14	Missouri................	3,582,111	....	....	....	....	6,990,711
15	Iowa....................	5,937,350	....	....	....	....	5,937,350
16	South Dakota............	....	....	....	5,800,000	....	5,800,000
17	Wyoming.................	5,656,509	....	....	....	....	5,656,509
18	Indiana.................	5,542,402	....	....	....	....	5,542,402
19	Alaska..................	12,282	....	....	5,125,000	163,845	5,301,127
20	Maryland................	4,318,211	....	....	....	....	4,318,211
21	Kentucky................	3,811,697	....	....	....	....	3,811,697
22	Nevada..................	....	....	....	2,371,882	....	2,371,882
23	New Mexico..............	1,600,588	....	....	....	....	1,600,588
	Total U. S., including remaining States...	$276,147,056	$248,577,829	$102,372,291	$70,096,021	$34,036,168	$750,680,827

A classification of iron ore production can not be made, further than that southern and western States produced $62,144,458, as against the remainder, which was all produced in the Lake Superior district. To Michigan is credited the entire Lake Superior output of this mineral, though, doubtless, a portion should be allotted to Minnesota and Wisconsin. This fact, however, could in nowise materially change the positions of respective States as shown above.

Lead and zinc are not shown, the products from these being of much smaller figures and of no bearing on this table, the totals in each case being credited to the States included in the table.

HALLOWED DAYS.

FIVE miles east of the present city of Butte rises the extreme apex of the eastern and western watersheds of the Rocky Mountains. Waters governed by the levels thus established start upon their widely separating courses, those descending the western slope following their devious ways — under the successive names of Silver Bow creek, and Deer Lodge, Missoula, Flathead, Pen d'Oreille, Clarke's Fork and Semiacquitaine rivers—into the Willamette river, below Portland, Oregon, and thence to the Pacific Ocean; while the waters descending the eastern slope, in like manner, under different names of creeks and rivers, finally complete their flow at the junction of the Mississippi with the Gulf of Mexico. To Silver Bow creek belongs the distinction of being the stream whose rise is further east than that of any other stream whose waters eventually reach the Pacific.

CONTINENTAL DIVIDE.
Elevation, 6,350 feet.

About twelve miles southwest from the apex or watershed divide and at a point where the waters of this creek have ceased their precipitous flow and have entered into the level of the valley, with an altitude of 5,700 feet above sea level, rests

HON. GRANVILLE STUART.

today the ruins of Silver Bow village, a drowsy relic of its former boom days.

In the summer of 1864, four prospectors — Budd Parker, P. Allison, and Joseph and James Esler — unmindful of the rich discoveries the previous year in Alder Gulch, at Virginia City — left that camp and pushed on across the main range of the Rockies, striking alluring placers along the banks of Silver Bow creek. It is worthy of notice that Silver Bow village ranks with Helena, or, as it was then more familiarly known, Last Chance Gulch, in point of discovery, as one of the pioneer mining settlements of the State, though never at any time so rich in placer deposits. Bannack was easily the pioneer of them all, followed closely by Virginia City, they being located in 1862 and 1863, respectively.

The point selected by these prospectors is "upon a bend of the stream, which forms a perfect figure of a gracefully curved Indian bow, and, from the mountain peaks which surround the valley, the glistening waters of the 'silver bow' etched in a shimmering sheen upon a dark ground of furzy grass, form a striking feature of the landscape." Thus was born the name of Silver Bow, which name was given to both village and creek.

While the advent of these adventurous prospectors marked the beginning of mining activity in this district, it is related that a party of prospectors, headed by Caleb E. Irvine, traveling through the section as early as 1856, found evidences of prehistoric mining.

News of rich strikes soon communicated to Alder Gulch and speedily a stampede set in and, like all mining-camps of easily opened and productive placers, the section sprang up rapidly. Prospecting was extended along the creek in either direction and, during the winter following, or 1864-'65, had proceeded to within six miles of its mouth — within the present site of Butte, on Town Gulch. This same winter several wooden structures were erected at Silver Bow and one store was erected in Butte.

By C. M. Russell, the Cowboy "Artist." AMBUSH OF PACK TRAIN. Reproduction courtesy of the Montana Historical Society

On February 6, 1865, a commission, of which Hon. Granville Stuart was a member, was empowered to lay out the town of Silver Bow. Silver Bow was made the county seat of Deer Lodge County in this year and on July 10 the first court was held. The village also enjoyed for a short time the distinction of being the capital of the young territory, but was soon removed by no other warrant than physical force to the village of Deer Lodge.

In the latter part of June of the same year the Democrats held at Silver Bow the first political convention of Deer Lodge County, and at the first election, upon September 1 following, the county seat followed the capital to Deer Lodge village.

By 1866 the entire creek channel from Silver Bow to Butte was worked by a company of four or more men to every two-hundred-foot claim. These toilers lived almost exclusively in tents or brush shanties adjoining their labors, worked faithfully six days of the week and generally showed up in one of the two towns on the seventh. This was the business day of the week. Gambling flourished, the merchant then made his weekly clean-up and the dance-house keeper panned out more than the richest placer. Prosperity was universal; every one was employed. Wages were $6 to $7 per day.

One writer at this time describes the style of architecture of the two towns as follows: "We should judge the prevailing style of architecture to be the Pan-Doric — a heathenish one of many evils. The material used is wood. Speaking of buildings, in Butte and Silver Bow, seven miles apart, year about houses are torn down in one and removed to the other. Last year houses were hauled from Silver Bow to Butte; this year the movement is reversed," concluding sarcastically: "This was to save timber, we suppose, as there is not more than a million or two acres of good timber in this immediate vicinity."

A decline in mining activity began in this vicinity in 1870 and even the revival of 1874-'75 did not strike the pioneer village of the county, and in 1880 the population had so dwindled that the census enumerators made no mention of the historic camp.

The early-day history of Silver Bow and that of Butte, which follows, is replete with the names of men who, at one time or another, became prominent in the affairs of the State. A great many have crossed the "Great Divide," while others have drifted to other parts in search of new discoveries. A few are still alive, some of whose names and faces have gone beyond the borders of the State and are found in the larger affairs of the nation. They were men who came to Montana, as did hundreds of others, by ox-team and on horseback, blazing the trail through an untraveled wilderness — over snow-clad mountains, across treacherous, unbridged streams and through valleys and passes infested with unfriendly tribes of savage Indians. Men who bore the hardships of the miner's life and discomforts of the primitive shack; who harbored their treasure, profited by frontier conditions and assured for themselves futures of plenty and comfort, and in many cases, of gigantic wealth, or, yielding to the lax moral conditions of the mining-camp, squandered their all in riotous living, and, in no few cases, are public charges today upon the charity of the city whose future they in part made possible. It is a story of a race for all — the survival of the fittest.

Hon. Granville Stuart, whose very faithful portrait is shown on a preceding page, is at present a most honored citizen of the city of Butte. Mr. Stuart antedates any living pioneer of the Silver Bow district, if not of the entire State, having, in company with his brother and a party of prospectors, passed through the section in 1858. Mr. Stuart, after having held responsible positions in municipal and State affairs at various times during his long residence in the State, more recently represented the Government as general consul to the Argentina Republic under President Cleveland, with distinguished ability. His recollections of early days are very vivid,

SILVER BOW VILLAGE.
"A relic of bygone days."

many of them having been reduced to print, and are worthy of careful perusal.

Meanwhile, in 1864, the same year of the original discoveries at Silver Bow village, William Allison, Jr., and G. O. Humphreys had pushed on up the stream and pitched their camp at the present site of Butte. At the time of their advent there were no stakes nor signs of mining having been previously prosecuted, save on what is known as the Original lode, where a hole four or five feet in depth was found. Indications pointed to the hole having been dug years before — by whom will probably never be known. No doubt it is the same hole reported to have been discovered by Caleb E. Irvine in 1856, and in all likelihood is attributable to the work of the native Indian.

Hon. Granville Stuart and others most intimately acquainted with early-day history are authority for the statement that the valley to the east and south of the new camp and running west to the Deer Lodge valley was the scene of much large game before the advent of the white man. Countless buffalo here found excellent grazing and were hunted by the various tribes of Indians adjacent to the region. It is likewise learned that many conflicts arose between these several tribes as to which should enjoy the supreme right to these hunting-grounds and many a hapless band of braves, separated from the main tribe by premature snows filling the passes of the divide, felt the sharp sting of chastisement for their presumptuous trespass.

"BIG BUTTE."

The initial settlers above mentioned were shortly followed by Dennis Leary and H. H. Porter. Rich placers were rapidly uncovered and a marked influx of goldseekers from Silver Bow and Alder Gulch resulted. So important were the discoveries and large the influx that in this year the first mining district was formed and the old town was located on Town Gulch and the name of Butte was given it. This name was derived from the majestic butte which reared its peak to the northwest of the new mining-camp, like a grim and lonely sentinel guarding the approach to the encircled valley within, rich in that vast treasure of mineral stores, the extent of which to this day — thirty-six years hence — has not been compassed.

In this year G. W. Newkirk, coming on from Alder Gulch, joined with Dennis Leary, T. C. Porter and the Humphrey Brothers in the erection of the first wooden house within the town, located on what is now Quartz street, and, until recently destroyed by fire, was occupied as a portion of the Girton house.

Even at this early date quartz-mining was receiving some attention, the first lead of this nature probably being the old Deer Lodge mine, now the Black Chief, this lead having been discovered by Charles Murphy and others in 1864.

The next authentic record of quartz-mining of an important nature is not found until 1867, at which time "Joe" Ramsdell, known, of all men, as the father of quartz-mining in this camp, struck a good character of ore in the Parrot lode and a company composed of himself, W. J. Parks, Dennis Leary, T. C. Porter and others was formed. A small arastra smelter was subsequently built by Charles E. Savage to handle the silver ore from this mine, but, owing to the insurmountable obstacles encountered, it was abandoned and all traces of the smelter almost immediately disappeared. Some ore from this mine was also taken to Swansea, but the enterprise presented too many drawbacks and active work was soon abandoned.

By C. M. Russell, "The Cowboy Artist." FOR SUPREMACY. Reproduction courtesy of Amon Carter Museum, Fort Worth, Texas

In the meantime the placer operations along the creek and up a portion of Missoula Gulch to the west of the city were booming, as was also the camp of Rocker, situated midway between Butte and Silver Bow.

In 1865 and 1866 the moral character of the town was probably the most deplorable of its placer days. It is said that at this time no man was safe without a brace of revolvers in his belt and a bowie-knife tucked in his bootleg. No small percentage of the numbers who had flocked to the district were of that daring, lawless type whose greatest pleasure was found in pastimes similar to "shooting up the town," which type has given to the entire West a name of "wild and woolly," and which name to this day has not been wholly effaced, though cause therefor has long years since disappeared.

This element, immune from the rapid methods of apprehension common to well-settled communities, driven from one section, found perfect safety for short sojourns within the confines of others, and it was of this class that the population of Butte at this time was in no small measure composed. The conditions thus erected were, no doubt, in a measure responsible for the establishment of the Mountain Code, which obtained throughout the territory about this time, and for the issuance in the year 1865 of an edict from the highest tribunal of this peculiar court — the territorial Vigilance Committee — serving notice upon wrong-doers of swift punishment wherever apprehended. That this notice was something other than a mere formality all old-timers will gladly testify, and early-day history abounds with accounts of many a hard-fought battle between outlaws and vigilantes.

The year 1866 was, all things considered, the most prosperous experienced by the camp as a placer-producer, and marked the advent of several settlers who afterward acquired a wide reputation, some of whom today are among the most substantial citizens of Butte, possessed of large holdings of real and mining property. Prominent among these settlers were A. W. Barnard, John Noyes, William Owsley, W. L. Farlin, W. J. Parks, A. J. Davis, David N. Upton and others. Mr. Upton, writing of Butte at this time, says: "There were no buildings where the town site is now located, but in Buffalo Gulch, near the present site of Centerville, there were about forty men and five women engaged in placer-mining with rockers who were doing pretty well."

By C. M. Russell, "The Cowboy Artist." "SHOOTING UP THE TOWN." Printed by kind permission of Schatzlein Paint Co.

During this year, however, and the succeeding one appeared signs only too visible to all, which cast a most forlorn horoscope for the future of the camp. Placer claims had about reached the climax of their productiveness and new fields were less frequently found and of inferior richness, and the spirit of uneasiness stalked unfeelingly about the camp. On the other hand, though there were many who had never faltered in their confidence in rich quartz deposits and who religiously picked away for unfound leads or who uncovered ores impossible of treatment by the limited processes known to the camp, there had not been one instance up to this time of a quartz strike which promised even a meager reward to its persevering owners and the hopelessness of failing confidence well nigh completed the desolation felt at the placer outlook.

Notwithstanding this overcasting prospect and subsequent events, the demand for schooling pressed itself upon the attention of the more serious minded of the camp, and in the winter of 1866-'67 the first school of Butte was established and was taught by Colonel Wood. Its life was short, but in the following winter a second one was opened, and since that time there has been at least one term of school each year.

The decadence of Butte as a placer camp, which began in 1866, became most pronounced the succeeding year, and before the close of this year the placers had given out completely. It was a blow that almost without exception has tolled the doom of every camp, which, prior or subsequent thereto, has owed its existence to placer-mining. Nearly every one left the district, disposing of their belongings to the stalwart few remaining, and each succeeding year for eight long years painted a more gloomy picture.

Notwithstanding the crushing experience thus sustained, there were a few whose confidence required a still further shock before their faith in Butte could be destroyed, for in this year the town site was laid out.

Now follows a dreary repetition of heroic efforts and almost invariable failure, the only surviving works being a mill erected by Harvey Bay, Jr., and Charles Hendrie, in 1868, and later known as the Davis mill. An important failure the same year was that of a furnace erected by Dennis Leary and Porter Brothers for the purpose of smelting ores from the Parrot lode. A bellows was used for a blast, but, ignorant of proper methods by which to flux the ores, they were compelled to abandon the effort.

By C. M. Russell, "The Cowboy Artist." SURRENDER OF THE OUTLAWS. Printed by kind permission of Schatzlein Paint Co.

The succeeding years of 1869 to 1874, inclusive, were uneventful ones, each one emphasizing a little more the downward progress of the camp, although representation work was unwaveringly performed upon the Parrot, Original, Gray Eagle and Mountain Brilliant claims and many others.

Of this period chapters might be written dealing with incidents that illustrate the tenacity with which the remaining little band of heroes clung to the latent possibilities of the camp as a quartz-producer. Pathetic experiences, culminating invariably in blasted hopes, were both the woof and the warp of the whole fabric of life during each succeeding day for these many long years. And yet the little handful that now composed the camp hung grimly on.

The labors of William J. Parks through these dark days are characteristic of the unyielding perseverance of these few. This tireless man, almost single-handed and alone, commenced work on his claim on the Parrot lode, alternately working for a short time for wages with which to secure a "grub-stake" and then returning to the mine and continuing his labors until his resources were exhausted. By this persistent policy, when at a depth of 155 feet, his labors and sacrifices were rewarded by the striking of pay ore. While thus toiling single-handed, it is related that a few well-to-do owners of claims on the same lode stood idly by and left it to this one man to develop the wealth of the whole lode. Thomas C. Porter, Dennis Leary and Henry H. Porter also ventured their all in developing the camp, coming from distant points yearly to represent their claims.

A particularly affecting story is told of James Gilchrist, one of nature's noblemen, beloved and honored by his contemporaries, who, after sinking shafts on the Original, Gambetta and Colusa lodes, was about to realize his dreams of wealth when his health completely failed him and he was forced to return to the East, where he soon died.

And thus might be written in different keys the experiences of nearly every individual who went to make up the little band of toilers. A few others who stand out conspicuously as having possessed an unshaken faith in the future of the camp were Capt. Nick Wall, William Berkins, Joseph Townsend, Capt. J. H. Rodgers and A. J. Davis, these members of the small roll of honor, year by year, performing their necessary representation work despite all sacrifices and all hardships. The efforts of all these early prospectors were directed toward the quest of quartz ores, but the old question of insufficient fluxes rendered all their labors alike unprofitable.

Copyrighted. Printed by kind permission of Mrs. Simon Hauswirth. BUTTE IN 1875.

There is a strange coincidence between this particular portion of the history of Butte, however, and that of many other camps and even of many individual mines of whatsoever section. While hopes had reached the lowest ebb and the future seemed most barren, there was hidden away in the uncovered levels of events, toward which Father Time was daily hastening, an epoch which was to revolutionize the whole future of the camp and proclaim it the greatest of its kind the world over. Some time previous to the date reached by this sketch, William Farlin, one of the early arrivals during the prosperous placer period of the camp, had left the district for other parts, taking with him specimens of the ores removed from several of the quartz leads about the camp. Journeying to Owyhee, Idaho, he had these specimens subjected to assay analysis and found that they were rich in the precious metals and carried some copper. He also acquired some beneficial information concerning the treatment of these ores.

Returning to Butte in the year 1874 he wisely retained his secret, plying his efforts ostensibly in the quest of copper ore, and patiently awaited the arrival of January 1 of the succeeding year, at which time became operative the new Congressional act relative to the forfeiture to the United States of unrepresented claims. At twelve o'clock on the night of the last day of 1874 Farlin placed notices of relocation on the Trevona and other lodes since made famous, these lodes falling under the conditions of the new act. By this action Farlin may be numbered as one of the first, if not the original, pioneer in a practice which has since become a universal one in mining sections, namely, the jumping of claims, the law going into effect at that time being practically the same as the one in existence at present.

WAITING FOR THE STAGE COACH.

The illustration herewith shown of "Butte in 1875" gives a most satisfactory idea of its general character at that period. In it will be seen the buildings shown in another illustration entitled "Waiting for the Stage Coach," which buildings at that time were probably the most pretentious of the city. The postoffice is shown at the left of the three structures in the larger picture; Simon Hauswirth's hotel — the only one at the time — occupies the center, and a saloon the one to the right. It was located on the corner now occupied by Clark Brothers' bank — Main and Broadway. Main street runs directly across the center of the picture from right to left. Those acquainted with present Butte will be able to trace out the sites of many present landmarks. A short distance to the right from the structures mentioned, along Main street, will be found where Granite street now crosses Main and upon respective corners of which are at present located the mammoth stores of the D. J. Hennessy Mercantile Company and the M. J. Connell Company. Tracing this imaginary street toward the bottom of the picture, running to the left, may be found the present site of the Noyes homestead. Continuing up Main street to the right, at its extreme end, toward the outer edge of the picture, will be found the Chastine Humphrey home — a white structure — showing the only tree which existed within the town site at that time. It is by the kindness of Mrs. Simon Hauswirth that we are permitted to present these two rare pictures, copyrighted copies of which are in her possession.

Now followed an exposure of the true facts concerning the value of the ores of the camp and the news spread like wildfire far and near, and newcomers flocked in in great number. Old locations were renewed and new ones made in rapid succession. The discovery of the Alice, La Plata, Burlington, Late Acquisition, Great Republic and other less famous mines followed quickly, and the movement toward Butte resolved itself into a stampede.

The town had become metamorphosed. From the hopeless, abandoned camp of a year before, it was now the Mecca of all who could possibly reach it and its growth was magic-like. The character of the new arrivals was a marked improvement upon a large portion of those attracted to the camp during placer days. The permanency guaranteed the camp by quartz-mining encouraged many to bring their families with them, and the town took on an air of stability theretofore unknown. Smelters for the proper handling of the various ores of the camp were begun in this year, the Centennial and Dexter being especially notable among them.

MISSOULA GULCH IN 1885.

Printed by kind permission of A. W. Barnard.

Following this period is a story of a steady growth and development in all directions. In 1878 the erection of smelters was being prosecuted vigorously. In this year a postoffice was created. By 1880 the population of the camp had reached 3,000. In the following year, under authority of a legislative act, the southern townships of Deer Lodge County were detached and organized under the name of Silver Bow.

On December 21 of this year was witnessed an event quite as pregnant with promise of a greater future for the camp as was the discovery of quartz. At 11 o'clock P. M. of that day the Utah & Northern connected Butte with Ogden, Utah, by a narrow-gauge railroad, which since has been widened into a standard gauge and has been a most potent element in the development of the city.

Mining was now being conducted over a large range of territory. In the Travona district, to the southwest, numerous mines were producing good returns and the Centennial and Dexter mills in the same district were being worked to their full capacity. The presence of water near the surface, however, made efforts in this locality difficult.

Running in a general way from north to southeast of the camp, with jagged spurs and erratic dips, was a foothill of commanding proportions, which common consent had dubbed "the hill." A mile or more to the north of the town proper, on the western slope of the "hill," the little town of Walkerville had sprung into existence upon the discovery of promising silver properties at this point, and at this time the Alice and Lexington had developed into enormous proportions, exceeding the wildest hopes of their owners. These easily were the most promising mines at this period.

Scattered on either side along the hill in its southeasterly course from Walkerville, until hill and valley merge into one, some two miles distant, were numerous mines of good promise. The Original, Parrot, Clark's Colusa, Ramsdell's Parrot, Mountain Consolidated and numerous other less noted mines were obtaining most satisfactory results. Smelters had been erected at convenient places over this large area to handle the ores from these many mines. Silver at this time was the metal exclusively sought, due to the presence of such large quantities of this character of ore in the Walkerville properties and the overwhelming proportion of silver found in the ores being mined in the other properties. Copper was encountered in no great quantity except in one instance. This exception was Clark's Colusa. In the early 70's W. A. Clark shipped a carload of ore from this mine carrying over thirty-five per cent of copper, to Baltimore, Md., by means of wagon trains to Corinne, Utah, and from thence by rail, but the excessive expense entailed prevented further shipments and work on the mine was discontinued.

Montana people generally will be pleased to find the accompanying very excellent likeness of Charles T. Meader, who did so much at this period to further mining in this section. By some he is yet known as the true father of copper-mining in the whole West, and is a fine type of the early settler. He was one of the original "forty-niners" to go to California by way of Cape Horn, and as early as 1865 erected a copper blast furnace in Calaveras County of that State, shipping the matte to Swansea, Wales, for final treatment. Mr. Meader came to Butte in 1876, purchasing the then undeveloped East and West Colusa claims. In 1881 he erected the Bell smelter. It was for Mr. Meader that the present suburb of Meaderville was named. Mr. Meader is at present eighty-two years of age and is located at Chewelah, Washington, where he is engaged in the pursuit of his old love, that of mining.

CHARLES T. MEADER.

The following year, or 1882, will ever stand as one of the great landmarks in the record of events. In that year occurred the discovery of the great copper body of the Anaconda mine. Its effect was revolutionary and it was this event which finally and completely established the permanency of the camp. The peculiarities of the ores of the Butte section had utterly failed to prepare the most visionary mind for such a wonderful deposit of the red metal and the discovery came as a tremendous surprise to all alike.

The advent of the railroad the year previous had removed all obstacles theretofore presented, and with the revelation that underlying all the mines operating along the "hill," outside of the Walkerville district, was an enormous deposit of copper, came Butte's second transition to a camp of a new character, which doubled and trebled the importance of the previous one, and old scenes were reënacted upon a larger scale.

Other properties which had been working on a reduced scale or had closed down, lying adjacent to the Anaconda, renewed their efforts with great vigor and, with each succeeding successful strike, there gradually dawned the truth that "the hill" was a veritable mountain of copper. Both the western and eastern slopes of the hill were now subjected to more careful scrutiny, and many mines sprang into existence. At the eastern extremity of the hill, as it descends into the valley and disappears, had sprung up the town of Meaderville. Almost without exception it was discovered that in the mines of "the hill" proper, or that portion lying south of Walkerville, the surface ores were richer in silver, but, as depth was gained and the water level passed, their character was changed overwhelmingly to copper.

In 1883 was emphasized the great importance to Butte of the copper discoveries in the Anaconda. While in 1882 the entire camp produced 12,093,750 fine ounces of gold, 2,699,296.38 fine ounces of silver and 9,058,284 fine pounds of copper, in the succeeding year gold gained but about twenty-five per cent, silver a trifle less than that percentage, and copper gained over 250 per cent.

The year 1884 was marked by no great incident save the increasing mining activity. The payroll of the mines and smelters for that year aggregated $620,000, with the Anaconda contributing $150,000, the Montana $65,000, the Lexington $50,000, and the Alice $50,000.

The following year was equally uneventful, unless that it more thoroughly established the preëminence of copper. The assessed valuation of the city at this time was about $7,500,000 as against $4,106,767 in 1881, and the gains in all directions of public growth over the latter year had been tremendous. From a turbulent, unsettled population, Butte had developed into a well-established city of 14,000, possessed of all manner of civilizing influences. The character of the buildings had increased with the growth of the town, and at this time many substantial structures of brick and stone had been erected, and many more were in course of construction. All lines of business had been introduced and from this time forward the garb of a typical mining town was gradually laid aside.

As a historical fact, it should be recorded that at this period placer mining was revived by the hydraulic process along Missoula Gulch. The gulch parallels Main street about one-third of a mile to the west, and from the accompanying illustration it will be seen how little the city had progressed in this direction at that time.

The principal mines of this period were the Anaconda, Original, Parrot, Colusa Parrot, Ramsdell Parrot, Bell, Mountain Consolidated, St. Lawrence, Mountain View and Colusa — all copper mines — and the Alice, Lexington and Moulton — exclusively silver.

GREATER BUTTE.

STEPPING forward to the dawn of the twentieth century, one stands amazed at Butte's wonderful development. Where, in 1885, rested a thriving mining camp of 14,000 souls — though even then recognized as the greatest of all mining camps on earth — there now stands a metropolis. Like an engulfing wave, progress and growth have placed their mark upon every nook of the city and entered into every cranny of its environments.

A population of about 65,000 people at present, or a 50,000 increase in fifteen years, tells its own story of growth in point of number. This means a logical growth in the city's limits. Where were once the humble shacks of the early settlers, there now remains but an occasional grim ruin, like a mocking skull, to conjure up the humanities of former days. In their place commodious streets, flanked on either side by business blocks and residences of the most modern types, disregarding in most instances the cowpath irregularity of hallowed days, cross each other at uniform intervals and run far into the valley on the south, to the hills on the west or the mining suburbs to the north and east.

From Walkerville south to Centerville, and from thence to Butte proper, one now passes as through one city. Meaderville, on the east, is rapidly being absorbed by the greater city. To the west, past Missoula Gulch and to the very base of " Big Butte," the city has pushed itself, and a mile south of the old town South Butte has been added to the city's suburbs and is as of the city itself.

Beautiful residences are the rule in new construction. Handsome church edifices and school structures are seen in every portion of the city and their influence is percolating the whole public mind.

AN OCCASIONAL GRIM RUIN.

Main street, once the sole thoroughfare, is the main street still, but what a change. Starting

REJUVENATED EARLY-DAY ARCHITECTURE.

from the southern limits of Centerville, it runs well nigh to South Butte, the greater portion of it rebuilt with modern blocks. Paralleling Main street on the east and west for varying distances are a half dozen less important business thoroughfares, while these, in turn, are crossed from east to west by as many more, equally as important as and vieing with Main for supremacy, far exceeding in architectural appearance the main thoroughfares of much older and more pretentious cities of the East.

It is excessively mild to say that no city in the whole West can boast of such scenes of bustling, crowding humanity as congest the main channels of trade from early morning until far into the night as may be seen in Butte on any day of the week. It is the marvel of every stranger and the result of Butte's wonderful growth — a growth which has received added significance in the past year by an increase of over 6,000 in population. A happy commentary, by way of an aside, is that the city has absorbed this tremendous influx with no apparent effort, and there is probably a smaller number of unemployed per capita among those who would work than any city in the country.

With the growth of the city has increased the morale of the people. Elevating influences everywhere have deeply implanted themselves and are rapidly becoming a powerful agency for good. The time has long passed when the license of the saloon and the variety playhouse found approbation in the best public mind. Though still in evidence, the latter has found its level in the lower portions of the city, frequented

WALKERVILLE.
Moulton and Alice Mines and Smelters in background.

only by a degenerate class common to all like communities, while the saloon boasts of as high a character as that traffic can boast wheresoever.

When it is considered that the personnel of Butte's population is in part made up of a floating element, gathered from all the nations of the earth, some of whom find employment in the mines and are colonized in cheap boarding-houses, unrestrained by elevating influences, and many never having known such influences in their whole lifetime, and having for their environment a city whose earlier antecedents were the manifold licenses of every new border town, the wonder is that there is any public morality. Yet, under such strenuous conditions, a public morality, stable and sure, is most certainly extending its leaven through the minds of the community.

CENTERVILLE.

Eloquent testimony to the improved morale of the labor employed in the mines is found in the rapidly increasing number of homes being erected by this class. Pretty little cottages of four and five rooms, built of wood and brick, are being

MEADERVILLE, LOOKING NORTHWEST.

East slope of "hill."

erected in every portion of the city. In most cases these are convenient to the place of employment, although the instinct of thrift and investment, taught by the successes of earlier settlers, has inspired many to seek the best portions of the city in the hopes of future enhancement of property. Oftentimes a double house is built, the rentals of one assisting the thrifty home-builder in the payment for both.

Many pungent object lessons are found throughout the city in structures of different periods belonging to one owner, showing a modest beginning in an old log shack, an expanding means in a more pretentious cottage, and the final attainment of an independent state in a pretty home, a business block or other like material evidences. With the growth of such instincts, a growth in other elevating attributes logically follows. Cleanly home surroundings, a higher standard of home life, assimilation of better ideals and more prominent participation in the city's affairs all have their beneficial effect, lending a responsibility to the individual, which is leaving its imprint upon the mining class and raising it to the plane of loftier citizenship. To the very small minority now belongs the objectionable element and, with this condition obtaining, the influences springing therefrom will, of their own force, speedily work out a like condition for all.

The city at present is most pronouncedly cosmopolitan. Nothing of which another metropolis may boast, in the way of up-to-dateness, is here lacking. Its public institutions are models of their kind. Twenty-eight schools and annexes are distributed throughout the city and its suburbs, attended the current year by 6,307 pupils and employing 170 teachers. Both curriculum and structures are of the most modern type. Gymnasiums and manual training are features of the system, while training-classes for the teachers assure a constant introduction of new normal methods, and military organizations and school teams in many of the sports add the touch of completeness to an otherwise broad learning. For the first time in its history, Butte is but beginning to experience the benefits which must accrue to a public morality from the quickened impulses and higher ideals which are knit into the mind of the student of a city's free schools and, in turn, become the uplifting heritage of that city's economy. The untold benefits which this new condition assures must prove a potent element in Butte's future life, as each succeeding

LOOKING SOUTH FROM CENTER OF CITY.

generation injects a new morality into the place of the lingering old. Besides these institutions, opportunity for special instruction is found in the new School of Mines, and a half dozen or more private schools of different character, teaching music, languages, etc.

The city is amply supplied with churches of a high order, twenty-eight church organizations telling their own story of religious activity. A public library, erected at a cost of $100,000 and containing 25,000 volumes, but emphasizes the trend of public improvement. Plans are about completed for the erection of a new federal building at a cost of $250,000, to accommodate the growing requirements of the postoffice service, with its six sub-stations, as well as other government branches. A beautiful courthouse adds its quota to the enhancing architectural excellence of Butte's structures, and the city government is comfortably housed in an attractive city hall.

LOOKING WEST.
School of Mines and "Big Butte" in background, to which the city extends.

Clubs of every description and organized for every purpose are making their influence felt in the regeneration of social and literary conditions, the Montana club supplying a public necessity as a club home for professional and business men and a place of entertainment for the city's guests, and is a model in its furnishings.

The theater has felt the uplifting influence of succeeding events and is patronized by the most critical of publics, who both exact and are furnished with the best in opera and drama that is known to the American stage.

LOOKING NORTHWEST. ORIGINAL MINE.

Of newspapers there are many. The Butte *Miner* and the Anaconda *Standard,* the latter published in Anaconda, twenty-eight miles distant, but with headquarters in this city, are daily morning papers, enjoying an eighteen-hour leased wire service of the Associated Press, and are of a much higher order than cities of twice the size in the East generally know. The *Inter Mountain* is an afternoon paper published six days in the week, and supplies the same high class of newspaper excellence as do the morning publications, it also enjoying the leased wire service of the Associated Press. In addition to these are nine weeklies, devoted to different purposes, ranging from mining news to religious matters.

The city is provided with power, water and light supplies sufficient for a city four times its size. Besides the old system of water-works, which supplies the city proper, and the Walkerville water system, there recently has been completed a system which gives much promise to the city's future, when it shall have become the leading manufacturing center of the State, which its logical location and prominence assure it. Some thirty miles distant from the city the Big Hole river — an exceptionally sanitary mountain stream — has been dammed and the electric power generated by this dam has been brought into the city, solving for all time the question of sufficiency of power. The water of the Big Hole is also piped into the city and large reservoirs are being prepared for its reception, which, in due time, will be connected with the city's present supply.

LOOKING NORTH.
Centerville in middle background on crest of "hill."

Railroads connect the city with all sections of the State and the whole country as well. The pioneer line, the Utah and Northern, is now a part of the Oregon Short Line system, connecting the city with the South and tapping the

LOOKING NORTHEAST.
West slope of "hill."

Union and Central Pacifics at Ogden, and the Oregon Railroad and Navigation system in Oregon. The Northern Pacific penetrates the exact center of the State from east to west, giving Butte a direct line to the Pacific, and, with St. Paul connections, to the Atlantic, and is easily Montana's leading railroad. The Great Northern reaches the city from Great Falls over the Montana Central Railway and handles the great quantities of ore shipped daily to the smelters of that city from the Butte mines. In addition to these roads, general agencies of the Chicago Great Western, the Burlington, Missouri Pacific, Oregon Railroad and Navigation, Rio Grande Western and Union Pacific have been established in the city, thus making Butte the center of railroad activity in the Northwest, as well as of commerce and mining.

The city has police and fire departments of unusually high standards. The former numbers a force of some forty patrolmen and detectives, is equipped with patrol wagon and is well disciplined, while the latter is supplied with all modern apparatus for fighting fire and, divided by sub-stations throughout the city and suburbs, has a most enviable record in saving property from that element.

Within the last few years the system of pavements has been extended along all of the principal business streets of the city. The material used is granite blocks, scientifically laid, and an extreme longevity is assured. The system is being pushed in every direction and is now reaching toward the residence portions.

Rapid transit, too, has felt the hand of improvement. Electric lines now connect the city proper with every outlying district, including Meaderville, Centerville, South Butte, the West Side and the Columbia Gardens, while the current year will doubtless see Walkerville added to the list. A close schedule maintained on these many lines permits a free flow of traffic to any point and gives the city as a whole a system not enjoyed by many larger cities.

MAIN STREET, LOOKING SOUTH.

Electric lights are, of course, a conspicuous part of the city's public improvements, the system contemplating not only the lighting of every section of the city, but also its extremest environments.

Of hotels there are many of a high standard.

BROADWAY LOOKING EAST.

GRANITE STREET, LOCKING EAST.

The McDermott and Butte are first-class American and European hostelries, respectively, while a new home for the traveling wayfarer, in the Thornton hotel, is in course of construction, and promises to be one of the finest of its kind in the Northwest. Private apartment houses, connected with up-to-date cafés, are becoming popular with Butte's people who prefer the quiet which they afford, and many of these provide elegant homes for this class.

Both the telephone and the telegraph, it seems quite needless to say, have long since reached the perfection mark. The latter is taking on the last improvement necessary to make it thoroughly cosmopolitan, in the acquisition of the district messenger service, while the former has just completed the recent innovations of the East, to wit, the abandonment of the bell system and the introduction of the electric light for call purposes, and the installment of long-distance service for all subscribers in place of older instruments.

One word should be said of climatic conditions in Butte, which, in a general way, applies to all of Montana. It is a question upon which rests much ignorance, and which, dispelled, removes a potent deterrent to many prospective home-seekers from other States who might otherwise look Montanaward.

Without hesitation and without equivocation, it can not be too heartily emphasized that no State in the Union enjoys so healthful a climate as does Montana. Its altitude, while not being excessive in any case, removes it from the blighting effects common to lower sections and provides for it an atmosphere which is pure and sweet and beneficial to the last degree. No water remains upon the surface to gather infectious germs, but finds its way immediately to running streams which carry it entirely away. The extreme dryness of the air secured by the altitude renders it far less penetrating than is the case with damp-laden air, and it is no conceit to say that thirty degrees colder weather is felt less acutely than in sections of lower altitude. In other words, one is less conscious of what ought to be the bitter cold of thirty degrees below zero in Montana than he is of zero weather in lower sections. An erroneous impression should not be created by reason of this statement, for, if the thermometer ever reaches that point, it is never for more than twenty-four or forty-eight hours throughout an

PARK STREET, LOOKING WEST.

Blaine. Greeley. Garfield. High. Washington. Lincoln. Adams. Webster.

REPRESENTATIVE SCHOOL BUILDINGS.

entire winter, and many winters pass without that point being reached. Its snows, too, do not acquire the depth popularly attributed to them. What would be rains in lower altitudes are often snows here, but they are short-lived and, once fallen, do not remain long upon the ground, being dissipated by the warm winds from the Pacific, which bare the ground as one sleeps. The holidays have usually passed before the ground is covered by a snow which survives one day of sunshine.

Butte, as a center of mining industry, does not exhaust its virtues. The prestige which its great wealth of mineral affords makes it the logical center of capital, commerce, politics, and a leader of social life throughout the State. Its capital is doing much, not only in developing other sections of the State, but also the rapidly opening sections of contiguous States as well as of British Columbia. This fact has not robbed the city, however, of the blessings of this agency, and great institutions which are reaching out into the whole State and attracting the trade thereof to Butte have been established in all lines of commerce. Wholesale establishments of every character, as well as general agencies of large concerns of the East, are rapidly centering in the city. Manufacturing, too, is beginning to find encouragement by reason of new conditions throughout the State and many ventures have found in Butte the most logical site for their plants.

City Library. County Courthouse. City Hall.

CITY AND COUNTY INSTITUTIONS.

No more fruitful subject is presented to the progressive mind for consideration than are the manufacturing opportunities which exist, not only in Butte, but throughout the State. Although one of the greatest producers of raw materials of any State in the Union, Montana thus far has consented to the importation of a large portion of the materials necessary to satisfy the needs of its people, whether it be for the covering of the body, the feeding of mouths or for the purposes of construction, or for materials used in the larger industries of the State. In Butte alone $1,000,000 a month is paid in wages to the people working in the mines and smelters, to say nothing of the large sums paid out monthly to thousands of others elsewhere employed, there being a total of 24,000 wage-earners in the city. Yet the production of a large majority of the supplies necessary to meet their demands is left to others foreign to the State.

With the vast amounts of wool, of hides and pelts and of mineral products of all kinds which Montana produces and the unequaled opportunities offered agriculturally, there exist advantages of which but few States can boast, and the wonder is that the present shall reserve to coming generations the task and profit of utilizing the great bulk of these raw materials where they may be obtained the cheapest, content to let interests wholly uninterested in Montana's development grow rich upon opportunities offered its own people. Yet this question, as well as many

Mountain View Methodist. First Presbyterian. Trinity Methodist, Centerville.
First Methodist, South. Episcopal.
First Baptist. Christian.

TYPES OF CHURCHES.

others, will undoubtedly be met, as have the questions of the past. The city is too large and its industries, as well as those of the whole State, are too great for them to be held back by these questions.

For the city a great future is in store, and the prediction is well founded that four or five years hence will witness a city as easternlike, as modern, and of as high a public morality as the misguided East can expect or the most exacting demand. What it will be in point of number conjecture alone can prophesy and events prove. That it will approximate a total of 100,000 is a conservative statement. Such a total would be in harmony with its present ratio of increase and every sign would point to an abnormal increase over present ratios rather than a recession therefrom. History records but few parellels to Butte — with such unpromising nucleus, rapid transitions, undesirable hordes, marvelous growth, whose Genesis to Revelations is but as one chapter, covering one generation — yet today it stands the peer in wealth and prospects of any inland Western city.

Mining activity of wider range but increases the belief that the supply of mineral is limitless, and that, as time shall unfold, works of greater magnitude covering an immensely larger field will take the place of those which today seem incredibly great and point the truth that Butte is not yet out of its swaddlings in greatness nor opportunity.

Church. Hospital. CATHOLIC INSTITUTIONS. Parish Home. Parish School.

True, it is not a haven for the unemployed of a whole nation, and it is not for the purpose of attracting the unemployed to this particular city that these statements are made. More specifically is it intended to show the general conditions which exist in the whole West, and point the thought that what exists in Butte today will exist elsewhere tomorrow, and that these general conditions are an invitation to a sturdy class of people who would carve out for themselves a

Hennessy. Owsley. Connell.
Mantle. Silver Bow.

A FEW BUSINESS BLOCKS.

future in a land where large opportunities exist. No more uninviting spot in the world could be found for profligacy than in the great West. No more tempting a field is offered for sobriety, ability and business acumen. It has ever been the graveyard of the indolent and ne'er-do-well, the goal of success for energy, pluck and perseverance.

PRODUCERS OF WEALTH.

An unpardonable slight would mark our effort to reflect true conditions if prominent attention were not drawn to the miners and smeltermen as a class, as upon them rests no light mantle of honor. For they are primarily the ones whose efforts really make Butte great. They are the men whose lives are jeopardized in the perilous duty of wresting riches from the bowels of the earth and of refining it, and it is the wealth so produced by them upon which a few are enabled to mount to positions of great power and influence. While the raw material produced by their efforts is shipped to the East and converted into dividends for the mine owners, their wages, amounting to a million or more a month, goes to increase the prosperity of the city. No factor which enters into the life of Butte as a metropolis stands so conspicuously in relief as a guarantee for greater things than does the enormous payroll of the wage-earners, always excepting the life of the mines themselves, and in the case of the mines of Butte, years of increasing productiveness are assured for them.

Too little thought is apt to be the portion of this courageous, honest, hard-working class of men. Not only does the miner, who goes beneath the ground to delve in semi-darkness, subject himself to accidents of every description, any one of which may render him a cripple for life or lay him cold in death, with a large family dependent upon him for support, but it is equally true of the smelterman, whose task at all times is a most hazardous and a dangerous one.

From eight to ten hours a day is the miner engaged at his perilous task, with giant-powder and deep, cavernous shafts as his work-fellows, with not the best of ventilation, subjected to the dampness common to deep levels, and, in no few cases, working in deep mire by reason of the presence of water, perspiring like a stoker, with hollow, unnatural noises as the only sounds to reach his ear, and shot to the surface or dropped to the bottom, thousands of feet below, day after day, with only a cable or steel belt between him and certain death. Yet he goes cheerfully to his task that he may meet the responsibilities of life like a man, educate his children to lives of usefulness and provide for his family in a manner fitting and appropriate. He is a hero, every inch of him, and he heroically performs a daily task from which a strong man might well quail and shrink back in terror.

The smelterman, too, does not escape the hazard consequent upon the production of the "red metal." Though working in the bright light of day, or by artificial light, which makes his vocation less dangerous, his task is not for children or weak-hearted men. Dangers are his, every moment of the day, and one has but to spend the briefest of time in witnessing the progress of his efforts in order to lend assent to this statement. As the furnaces are tapped and throw out their myriads of sparks of molten matter, which mercilessly burn to the bone whenever human flesh is touched, and as the larger volume comes spluttering out, the smelterman stands closely at his post of duty, conveying the burning liquid to converter and back to furnace again, if necessity requires, subject at any moment to direful accidents which may horribly mutilate him or cost him his life. All honor is due to the class of men who will consent to follow a vocation so dangerous — an honor as lasting and as exalted as was ever extended to the pioneer.

No less than thirteen thousand men are employed in the mines and smelters. Their pay runs from $3.50 to $4 per day, and, in some cases, still higher, and a finer aggregation of wage-earners can not be found the length of the land. They are an extremely intelligent class, differing in this respect from those of many other mining sections, and among them will be found the student of every profession, the musician, the thinker, the mining expert, the orator and the political leader. Many of the leading professional and business men of the city are graduates from the mining and smelting classes, and their success denotes the character of the men engaged in these vocations.

T. M. Hodgens. Hon. W. A. Clark. C. W. Clark.
Mrs. P. A. Largey. J. H. Vivian.

PROMINENT RESIDENCES.

The nationalities represented among these two classes are principally American, English and Irish, although other nationalities are encountered in smaller numbers. The vast majority of the thousands following the respective vocations are thrifty, have built pretty homes, own bank accounts, attire their families in the warmest and best and are raising their children in such a way as to make them aggressive and independent, giving them good educations in the public schools and, in a large number of cases, providing them with higher advantages. Unlike mining sections of other States, no class distinction is drawn between those working in the mines in Butte and the general public, either socially or otherwise, it being recognized that some of the best types of citizens are thus employed.

The miners and smeltermen, as well as every other branch of labor in the city, have their labor organizations, which have done much to improve the conditions and morals of the entire class. One lodge alone of the Miner's Union numbers something over six thousand members, and the union as a whole is the parent organization of all organized mining labor throughout the West, the lodges or branches of the larger organization, until a short time ago, if not at present, receiving their charters from the Butte body. Many of the officers of this larger body are, or until recently were, employed in the mines of the city and are recognized as the foremost advocates of organized labor in the entire West.

MINERS' UNION HALL.

BOSTON AND MONTANA BAND.

To the miners of Butte has been reserved the proud distinction of furnishing the finest musical organization in the State. It seems quite unnecessary to state that this organization is the Boston and Montana Band, the entire State for many years having recognized this truth.

Prof. S. H. Treloar is the organizer of the band, which took place so long ago as December 22, 1887, and for these many years the band has had no successful competitor for first place.

The band was encouraged by a former manager of the Boston and Montana mines, Capt. Thomas Couch, himself an ardent admirer of high-class band music, and, as the original members, numbering six men, were employed by the Boston and Montana Company, the organization took the name of that company for its own. For over a year and a half after organization, the band did not make a public appearance, confining itself to careful study and gradually increasing its membership, until it numbered eighteen. Its first appearance was met with an enthusiastic welcome and its popularity slowly increased, until the whole State learned to seek its services when a high order of music was required. When it is remembered that many of the military posts had their crack musical organizations, the palm thus yielded to the Boston and Montana Band receives added significance.

In June, 1890, the band was installed as the regimental band for the Montana National Guard and its presence did much to enhance the interest taken in the State encampments of that military organization.

In May, 1892, the band was incorporated and at this time numbered twenty-one members. The band henceforth began to receive engagements from all parts of the State and was oftentimes taken beyond the borders. In 1896 it was taken to Chicago by the Montana delegation to be in attendance upon the national Democratic convention. The press notices of the Eastern papers

BOSTON AND MONTANA BAND.

Montana's finest musical organization.

were extremely flattering, much surprise being expressed that so high class a musical organization could be assembled so far West and among a class of men who went beneath ground to earn their daily bread. So great was the interest felt in the band throughout the East that, while returning, the band was met at the depot in Minneapolis by a large committee and made to leave the train and become the guests of the citizens of that city. The continued success of the band has never faltered and, including this year's engagements, it has realized about $60,000. The current year has been by far the most successful one of its existence, something over $12,000 having been turned into its treasury as a result of the year's engagements. The band now comprises about thirty members, who are well equipped with the very latest instruments. They own their own headquarters hall and are most zealous in their practice and technical study. The men are all practical miners, engaged in different duties about the mines, and are a fine body of men.

They were in attendance upon the national convention of the Democratic party at Kansas City the current year and to them was accorded the high honor of adding the keystone to an otherwise spectacular scene during that convention. American flags had been lowered over a portion of the stage during the proceedings of the convention and behind them had been placed an heroic-sized bust of Hon. W. J. Bryan. When the proper moment arrived and the flags were drawn aside amid tremendous excitement, it was the Boston and Montana Band that came down the center of the aisle playing "Dixie" as a fitting climax to the impressive scene.

The great success of the band lies largely in the fact that, while every one is an artist unto himself, their occupation is such as to more largely develop their lung capacity and thereby give greater zest and tone than is possible in many other musical organizations, its flattering prominence at Kansas City being largely due to this fact.

The band has had no little part in eliminating baneful class distinctions throughout the city, its entire personnel having long ago demonstrated that as high culture can work below ground as above it.

Professor Treloar is still at the head of the great organization, and this fact in itself is a standing promise of the perpetuation of the previous high standard of the band.

MINING MAGNATES.

Before passing to new subjects, a pause is mandatory for the review of a few of the men who, as actual residents of the city, have done the most to develop its wonderful resources by bringing it to its present high position and assuring for it a future still more exalted. Many persons might be brought under this head with the greatest propriety, did space but permit, and, in confining the number to a few, regret is felt that a fuller justice can not be done to scores who have builded so well and so unremittingly. What the city is today is the result of the united efforts of many, who, crowned by success, have, in turn, transmitted a measure of their success to the upbuilding of the whole community.

M. J. Connell, D. J. Hennessy, A. J. Davis (deceased), A. W. Barnard, William Owsley, John Noyes, Hon. Lee Mantle, William Thompson (deceased), and others, all have had their place in Butte's history, and all of these who survive are among the most prominent citizens and business men of the city to-day. Scores of others, in varying degrees of prominence, have been identified with the large things of Butte's development, and failure in their mention in nowise is a slight upon their worth. Other works extant deal most interestingly of these men. The nature of this effort precludes their treatment in a like manner.

Butte, and Montana as well, for many years has enjoyed the distinction of being the scene of the largest operations of two giants of national repute — William A. Clark and Marcus Daly. Each was supreme in his given field, Mr. Clark, as the largest individual mine owner in the world, and Mr. Daly, as easily, as the largest manager of mines. Their fields of action have never been the same, when carefully studied, although both were identified with a common object, the mining industry. As owner, Mr. Clark's activity tended always to the acquisition of properties as a personal investment. As manager, Mr. Daly was associated with co-owners for the purpose of developing their properties.

WILLIAM A. CLARK

Each, in his respective function, became a giant among men, and the life of either is a most fitting example of what an unswerving will and determination will do in the grappling with and bending of events as they present themselves in the course of human life. Opportunity they both had, but it was not so much the opportunity as the determination to grasp it which characterized their lives.

Both men started from the bottom rung and relied upon their own faculties and own resources to work out their careers. While both became immensely wealthy wholly by their own efforts, thus weaving an example for others who would emulate them, the more commendable fact remains that they also builded for themselves a character which, as citizen, as man of affairs, as husband, father and son, enabled them to reflect some of the noblest traits which mankind can own. Both have ever been men among men, counting their friends by the thousands and scattered through every State of the Union, though to Montana has fallen the heritage which their genius has wrought.

Perfections were never embodied in any human, and, if varying degrees of success have marked their lives, natural causes have worked them. Different temperaments, qualifications of mind and body widely at variance, opportunities springing from different events, environments, associations and the dozen other influences which are responsible for every life have all worked their part, and the memories of both will ever be reverenced by the people of Montana. Both were supernaturally endowed, and have easily proven themselves the superior in genius of their fellow creatures. Only as the older resident pioneer of the section treated, precedence, in point of order, is given to Mr. Clark.

WILLIAM A. CLARK.

The subject of this sketch was born near Connellsville, Fayette County, Pennsylvania, in 1839. His father was a farmer, and, while working summers on the farm, Mr. Clark attended the public schools during the winter months until fourteen years of age. He then entered Laurel Hill Academy for a short time. In 1856, he, with his father, moved to Iowa, where he worked on the farm for one season, teaching school the following winter. The following term he attended an academy at Birmingham, thereafter spending one term at the University at Mount Pleasant. Following this he began the practice of law at the latter place, which profession he prosecuted for two years. In 1859-60 he taught school in Central Missouri.

Two years later, or in 1862, was born the idea in Mr. Clark's mind to go West, and, when but twenty-three years of age, he mounted the seat of an overland wagon and started out into that great expanse in quest of fortune. The country was wild and dangerous, and "blazing the trail" presented obstacles sufficient to deter an older and stronger heart.

He arrived in South Park, Colorado, in due time, and for a year worked in the quartz mines of that place, little thinking that in this humble capacity he was laying the groundwork for a career which would eventually stamp him as the "largest individual mine-owner on earth." While thus engaged, stories of the discovery of large gold deposits on Grasshopper Creek, at Bannack, Montana, determined Mr. Clark to emigrate hence. After sixty-five days of severe hardships, traveling by slow-going ox-team, he arrived at Bannack on the 7th of July, 1863. The character of the man finds eloquent testimony in Mr. Clark's first act upon his arrival and reveals the secret of his repeated subsequent successes — his promptitude to act wisely on the instant. Though weary with travel, he joined a stampede party on the same night and secured a placer claim on Horse Prairie. The wisdom of the step succeeding events proved.

After working his claim until the close of the season, Mr. Clark found himself possessed of his first thousand dollars, which amount served as the basis for his present immense fortune, his wealth thenceforward being marked by a steady increase over this sum. With the coming of winter and the cessation of mining operations until spring, Mr. Clark's shrewd business instincts discovered opportunities for increasing his small means. Purchasing a mule, he set out for Salt Lake, where he stocked himself with much-needed provisions, which he sold on his return at extraordinarily high prices. In the spring he again resumed mining, continuing his operations until fall.

With his means thus enhanced by mining and trading, he now sought larger opportunities. Selling his mining interests at Bannack, he again returned to Salt Lake, where he purchased a larger supply of merchandise of a general character and freighted it to Virginia City, arriving there in the winter of 1864-65. It was during this winter that the great flour famine, which spread over the entire State, occurred, and flour sold at from $1 to $1.50 per pound. Riots ensued, all flour that could be found was seized, and many persons were compelled to live exclusively on meat and beans for a long time. While Mr. Clark did not hold his flour, as did many others, at prohibitive prices, he nevertheless received large legitimate profits by reason of its great scarcity.

From now on, for the next three years, Mr. Clark availed himself of his greatly enhanced means in conducting mercantile enterprises on a much larger scale. By bold freighting expeditions from distant points, some even so far as San Francisco, into sparsely settled sections where comforts were scarce, he rapidly accumulated a fortune of considerable size. In one tobacco transaction alone he netted many thousands.

By 1868 his enterprises covered a large range, including sub-contracting of mail routes, Mr. Clark at this time having Helena as his headquarters. In 1870 he entered into partnership with Mr. S. E. Larabie and others and a banking house was established at Deer Lodge by the firm, a general mercantile business also being conducted. The latter interests were disposed of in the summer of this year and exclusive attention was paid to banking by the firm. In 1872 they organized a national bank, Mr. Clark being elected its president. The purchase and shipment of gold dust formed a leading feature of the institution's transactions, this alone amounting each season to over a million dollars. In 1878 the charter was surrendered and the business was continued under the previous firm style. A branch at this time was also established at Butte. In 1884 Messrs. Clark and Larabie purchased the interests of the firm, and for some time afterward continued this partnership, though it was finally dissolved.

Meanwhile, as far back as 1872, Mr. Clark had turned his attention to quartz-mining in the Butte district, purchasing an interest in the Original, Colusa, Mountain Chief and Gambetta mines. In these acts is found a culmination of the benefits which he acquired many years previous in his quartz-mining experiences in Colorado. Not content, however, with his crude knowledge of this character of mining, gained from practical experience, Mr. Clark recognized the necessity of a more exact science, and, during the winter following, studied assaying at the School of Mines, Columbia College, New York City.

With the discoveries of the Trevona and other silver properties in 1875, Mr. Clark entered quite extensively into prospecting for and the purchasing of mines of this character. The Dexter smelter mentioned previously was completed with funds furnished by Mr. Clark, this smelter being the first successful silver stamp mill operated in the camp. In 1875 Mr. Clark located the Moulton mine at Walkerville, arranging with a syndicate to improve the same, including the sinking of an 800-foot shaft, at a total cost of $400,000. In May, 1879, he organized the Colorado and Montana Smelting Company.

By 1885 Mr. Clark was part or entire owner in no less than forty-six paying silver or copper properties, many of which have since been abandoned, while a number of others have developed into enormously rich copper mines.

Mr. Clark has from time to time relinquished his holdings in properties in which he held a minority interest, retaining those in which he was entire or major owner, thereby displaying one of his strongest characteristics — a desire to be independent in all of his investments. The same shrewd business acumen which had characterized his operations from the first quite naturally was displayed at such times as these holdings were released, and Mr. Clark never allowed any great time to elapse before utilizing the means thus secured in investments equally promising in returns. These investments covered every resource of the rapidly growing territory, and gradually extended over the most promising sections of the whole West, including not only mining, but the development of great plantations, such as coffee and sugar beet, and the construction of railroads.

His United Verde copper mine at Jerome, Arizona, is, perhaps, the richest mine of its kind in the world, producing an enormously high percentage of ore, which is mined literally by quarrying, without the counter expense of timbering, etc., incident to other mines of like mineral properties. He has also erected in the East large refining plants for the final treatment of ores from his many mines.

Mr. Clark, though an unusually active man in the conducting of his rapidly growing and diversified interests, has found time for the gratification of various other dominating instincts, artistic, scholastic, social and political activity all having their quota of time and thought. He is a gentleman almost delicate in appearance, refined and cultured, and capable of versatile conversation on subjects of wide range. Whether as a humble wage-earner, as a man of growing means and larger ideas or as a giant in the mining world, he has ever been the same frank, courteous gentleman, easy of approach, considerate of the feelings of others, and always ready to lend his generous aid and kindly counsel to movements which promised good for the State or for the people thereof.

Mr. Clark was married in 1869 to Miss Katherine Stauffer, of his native town, Connellsville, Pennsylvania, and, though deceased for many years, there still survive her a host of friends who never tire in extolling her beautiful character, sweet disposition and womanly loveliness. Six children were the fruits of this happy union, four of whom, now grown, still survive. Two sons, Charles W. and William A., Jr., reside at Butte, and are identified with many of the largest institutions of the city, conducting their own affairs and promising to follow in their father's footsteps as men of large ability and executive attainments.

Mr. Clark maintains a beautiful home in Butte, where he spends a large portion of his time. A beautiful residence is in course of erection on Fifth avenue, New York, if, indeed, it is not already completed, to be occupied by his two married daughters, who reside in that city. His mother is living and resides at Los Angeles, California, and is visited very frequently by Mr. Clark, who has surrounded her with every luxury in her old age. Mr. Clark is vigorous and intensely active, and Montana will doubtless enjoy many additional benefits from his generous hands, as in the past, before that unfortunate day when he shall join that large body of pioneers who, in proportion to their ability, have helped to work out a great destiny for the young commonwealth.

Mr. Clark is variously reputed to be worth from $50,000,000 to $100,000,000. Whether the first or last figure is more nearly correct Mr. Clark himself probably could not state, intrinsic and income valuations always differing so widely. A statement oftentimes made and never disputed, however, is that Mr. Clark's annual income ranges from $5,000,000 to $6,000,000, or five to six per cent on $100,000,000.

MARCUS DALY.

(DECEASED.)

In presenting a brief sketch of Mr. Daly's wonderful career, so closely identified with the large things of the whole State as it was, it becomes the sad duty to also chronicle its sad close. When a portion of the sketch which follows was first prepared, Mr. Daly was in full possession of all the powers which his wonderful genius had wrought for him, and, though suffering from the malady which finally brought his lamentable end, no one dreamed that the mortal coil had so nearly unwound. To him Montana has erected a monument more beautiful and more lasting than could be hewn from the granite which today is being extracted from the deep levels of the Great Anaconda — the masterpiece of his great genius. For an immortal column that reaches to heaven, bound with ties of love and affection, lingers in the minds of the people of this great commonwealth, whose memory runneth back over the score of years just ending, recalling his modest advent, his rapid ascent to places of power and wide influence, his love of and faith in the great commonwealth, his devotion to friends of every degree and walk of life and countless acts which will endear him in their memory, until the grave shall claim them, and whose children's children will ever preserve a tender place in their hearts for this great giant among men, who builded so well and died so nobly.

Of Mr. Daly's early life and many important facts throughout his whole career knowledge is lacking to a most lamentable extent. He was born in a hamlet on the edge of Ballyjamesduff, County Cavan, Ireland, in about the year 1841. He is supposed to have spent an uneventful boyhood upon the farm. Educational advantages, it is well known, were wholly denied him, and, when he secretly decided, at the age of fifteen, to seek new fortunes in America, he was reinforced by nothing that could possibly aid him in rising in the world save an indomitable will, which ever characterized him, and a strong physique. He first found employment in Brooklyn as a dockhand, but the work was severely hard for his young years and a desire for further travel soon possessed him.

Deciding to emigrate to San Francisco, he lived frugally and saved his means until he had accumulated sufficient to pay his passage by way of the Horn. He eventually arrived in the latter city

MARCUS DALY

in rather straitened circumstances, but in what year all record seems to be lacking. Here he found it difficult to secure employment, and for a time he earned a livelihood by various means, having no trade to assist him. His earlier farming experience, however, played him a good service here and he was enabled to obtain frequent employment as a farm or garden helper. At other times he secured subordinate positions about the adjacent placer camps. He finally obtained permanent occupation in the quartz mines of Utah, after having drifted between this State and California a number of times. It was during this period that Mr. Daly became acquainted with George Hearst, afterward the California millionaire, who at that time was an ardent prospector, though then possessed of little means. This acquaintance afterward proved of large benefit to him.

Mr. Daly continued in his position in the quartz mines until the year 1876, nothing of event transpiring, though he was naturally absorbing the principles of this character of mining and laying the foundation for a future which ripened him into its greatest exponent, perhaps, the world over. In this year he decided to cast his fortunes with the mining operations of the rejuvenated Butte. The previous year had occurred the discovery of the Travona and other promising properties, and Mr. Daly's advent was at a time when many others were stampeding in the same direction.

Many persons who are most familiar with Mr. Daly's life at this time assume that Mr. Daly came as the representative of the Walker Brothers of Salt Lake, and that he acquired an interest in the Alice mine, which was owned by them, as a part consideration in his assuming control of that property. Others say to the contrary, claiming that Mr. Daly arrived without means, but subsequent events would seem to disprove their contention. This counts for little, however, save to rob a most wonderful career of a little color of romance and to cloud the real event whereby he acquired the foundation for his future wealth. Suffice to say, Mr. Daly was soon in complete charge of the Alice mine, and it was Mr. Daly's tremendous success in this capacity that earned for him the recognition as the ablest practical mining man in the camp, as well as a leading expert throughout the whole West.

In another place it has been shown that the Alice soon became the richest silver property in the camp, having in its day produced many millions for its owners. It was here that Mr. Daly demonstrated his large ability as a developer of mining properties, displaying almost superhuman shrewdness and oftentimes proceeding against the unanimous counsel of the best experts of the camp, but invariably proving the correctness of his position. It is not too broad, perhaps, to state that, but for him, many of the adjacent silver properties, afterward made famous, would never have been developed on so enormous a scale but for the shrewd mining judgment displayed by Mr. Daly in establishing the value of the Alice lode.

But a larger field was destined for Mr. Daly. While engaged in the management of the Alice, other properties had been developing on a moderate scale, and among these was the Anaconda, which was producing a good quality of silver ore. Realizing that this new mine had a large future before it, he relinquished his connection with the Alice property for the purpose of identifying himself with the Anaconda. Here again arises the contention as to whether Mr. Daly was enabled to receive a cash consideration upon his severance of relations with the Walker brothers, but, as Mr. Daly was eventually a large holder of Anaconda stock, the assumption seems logical that he must have had means at this time with which to procure the same.

A deal was finally completed, however, through the agency of Mr. Hearst, whereby that gentleman and Mr. Daly, together with Messrs. Haggin and Tevis, became joint owners of the Anaconda property for the consideration of $30,000. Mr. Daly, by the terms of the deal, passed to the head of the mine in the management of its active operations, and shortly the strike was made which so wonderfully revolutionized the character of the camp from a silver to a copper one.

The demands for larger smelter accommodations now presented themselves, and Mr. Daly was given instructions to locate an available site for their erection. After considerable study of the situation, Mr. Daly finaly decided upon Anaconda — thirty miles distant — as the most practicable location, and work was begun upon the construction of smelters, which, in time, grew to be the largest, probably, of any smelting plant in the world. With the growth of not only the original Anaconda property, but also numerous other mines acquired by the company at Butte, including the St. Lawrence, the growth of the

PATRICK A. LARGEY

city of Anaconda kept pace, until, today, that city is one of the handsomest in the State, with a population of about 13,000.

To better facilitate the handling of ores from these many mines an independent railroad was constructed, known as the Butte, Anaconda & Pacific, which is one of the finest equipped systems to be found anywhere for both passenger and freight traffic. The demands upon the smelters at Anaconda grew to such gigantic proportions that a second immense smelting plant soon became necessary and, more recently, still another has been started, and is in course of erection at Anaconda, which will double the capacity of the old smelters. Mr. Daly, as its founder and most enthusiastic patron, was closely identified with everything which had for its object the improvement of Anaconda, erecting an elegant hotel, which is probably the finest in the State, establishing a newspaper of the very first class, and in every way bringing the city up to the highest standard.

The Anaconda mine, under Mr. Daly's shrewd management, soon grew into a colossal system, including many new mines lying adjacent to the original one, until today it is of tremendous proportions, employing thousands of men and having diversified interests all over the State.

Besides his interests in this property, Mr. Daly was responsible in great measure for the development of the great agricultural possibilities of the Bitter Root valley, in Western Montana, where he established his famous Bitter Root stock farm, as well as putting countless acres under cultivation, and today his stock and fruit ranch in that valley is one of the handsomest and most complete in the entire country. He was a great fancier of blooded racing stock and his colors have led the way in many of the large Derbys of the East for many years past. Mr. Daly was also interested as large stockholder in many of the leading mercantile establishments of both Butte and Anaconda, besides many manufacturing enterprises scattered throughout the State. No instances are known where he engaged in mining operations on an independent basis, although the Washoe Mining Company's stock was largely held by him. The constantly enhancing values of the great Anaconda properties, however, yielded him a princely fortune, to say nothing of his other immense holdings.

Mr. Daly has made his home at Anaconda in recent years, although spending a portion of his time in the East in consultation with other large stockholders of the Anaconda, and on his Bitter Root ranch with his family, who have made it their permanent headquarters for some time.

Mr. Daly was married at Salt Lake, in 1872, to Miss Margaret Evans, who survives him, and has a circle of friends larger than the borders of the State, who mourn with her in her dark days of affliction. Four children also survive Mr. Daly — Misses Margaret, Mary and Hattie, and a son, Marcus.

He was an enthusiastic lover of Montana, and had he been spared for further years would have been an invaluable agency in bringing forward the resources of the whole commonwealth, as he so loyally did in the past. Mr. Daly's fortune has been variously estimated at from $15,000,000 to $50,000,000, but it is believed that the first figure is more nearly correct, though the broadening of his opportunities and the working out of plans known to have been formulated by him previous to his death, in a few years would have added immensely to this amount.

To few men is given the privilege of gathering around him so many warm friends as Mr. Daly possessed, who saw in him the personification of many virtues, of which not the least was his kindly consideration of friends of early days, many of whose widows today have reason to bless his memory. Mr. Daly was an extremely modest and retiring gentleman, who aspired to no elective or appointive political preferment, and seemed happiest in the pursuit of his business duties or in the quiet of his home.

PATRICK A. LARGEY.

(DECEASED.)

When the present shall have become crystallized into the past and a more accurate view of events shall permit, few names will stand out in such relief as will that of the late Patrick A. Largey, in connection with Butte's development. Born of modest parentage, he took into life the sterling qualities of integrity and business ability, and with these wrought out for himself a handsome heritage, besides leaving behind him throughout that life — thirty-three years of which were spent in Montana — a path of kindly deeds and ennobling examples.

Mr. Largey was born in Perry County, Ohio, in 1848, and, as a young man, engaged in mercantile pursuits in Iowa. In 1865 he crossed the plains at the head of a wagon train of sixty wagons, having as a business associate John A. Creighton, at present a large property-holder in Montana and a resident of Omaha, Neb.

Arriving at Virginia City, Mr. Largey became engaged in merchandising and invested quite extensively in placer properties adjacent to that famous camp, which he operated for many years. He continued in business in Virginia City until 1879, and in 1881 adopted Butte as his future home. Meanwhile, Mr. Largey had broadened his field of activity and usefulness, and in the year 1878 constructed overland a telegraph line from Virginia City to Butte and hence, by way of the Deer Lodge valley, to Helena and Bozeman, thereby preceding the railroad by three years, and affording the isolated people of the State the first opportunity for rapid communication with the outside world. Hon. Lee Mantle was at this time engaged as telegraph operator in Mr. Largey's service, and, to his connection with Mr. Largey and the benefits arising therefrom, Mr. Mantle's later success in life is, doubtless, no little due.

Mr. Largey's interests developed rapidly and his entire fortune, as well as thought, were loaned to the city's development. Both mercantile and mining pursuits received his attention. In the former he was associated with many men who were at the head of different mines of the locality, and in this manner opportunity was offered him to secure valuable properties. It is said by many of his old associates that he was the possessor of more patented mining claims than any other man in the United States. The most valuable of his mining acquisitions was undoubtedly the Speculator mine, and the returns therefrom are supposed to have made him one of the largest operators in the State. With two other gentlemen, Mr. Largey established the first electric light plant in the city. He also established the Butte *Inter Mountain,* and for many years was the president of the company publishing that journal, and was stockholder in many other leading establishments of the city.

Mr. Largey's most useful and successful career was brought to a sad termination on January 11, 1898. Some three years previous had occurred the direful explosion of giant powder which resulted in the death of some three-score persons. Many others were injured and much ill-directed feeling was engendered thereby. Mr. Largey was least responsible for the explosion and its calamitous results than, possibly, any man in the city, yet he went far beyond the most exacting requirements of duty or public spirit in appeasing the suffering caused thereby. He was especially annoyed by one Thomas J. Riley, who repeatedly called upon and demanded assistance from Mr. Largey, after the latter, in connection with others, had contributed $1,500 to compensate Riley for the loss of a leg in the explosion. Riley's demands at last became threats and, while engaged in his duties as president of the State Savings bank upon the day mentioned, Mr. Largey was cruelly shot down by the former as a revenge for his injuries — injuries for which Mr. Largey never was in any way responsible.

Mr. Largey left a widow, formerly Miss Lulu Sellers of Chicago, to whom he was married in 1877, who accompanied him in those rugged early days wherever his interests took him, and who today is administering his affairs where his interrupted life laid them down. Four children also survived him. To the stricken family, however, was not confined the grief caused by his death. Not only the city, but the whole State recognized and mourned the loss of one of the best types of the men to whom all time must accord the honor of having achieved the development of the great West, as well as a man, who, with a few others, made Butte's future possible and attracted to Montana the causes and events, which, in their unwinding, are making and will continue to make it one of the greatest States of the Union.

F. AUGUSTUS HEINZE.

The last, though not least, of the prominent characters to whom special mention is due as one who has done much to develop Butte's wonderful resources and who, as one of the large owners of its mining properties, stands as joint sponsor for a still greater future, is F. Augustus Heinze.

Some eleven years ago Mr. Heinze came to the city as a mere boy — about twenty years of age. Of means or resources he had but little, so far as is known, although he had the advantage of the subjects of the preceding sketches in that he had received an advanced education in the very things necessary to successful mining operations. He was well versed in metallurgy, geology and the other essentials to mining, and to this learning was added the keenest of intellects and shrewdest of natures.

Mr. Heinze first entered the mining field in a moderately humble capacity — that of mining

F. AUGUSTUS HEINZE

engineer for the Boston and Montana Mining Company — but his quick perception soon discovered the larger possibilities which were presented to him, and he concluded to try for higher things. Before a year had passed he had a most thorough knowledge of the intricate mineral formations of the entire mining district, and was so sure that there was room for him in this great industry that he shortly returned to New York to facilitate matters. Even at this extremely youthful age, before a year had passed he had succeeded in enlisting capital sufficient to erect a large smelter at Butte for the purpose of reducing ores from some of the independent mines not provided with such plants.

A company was formed under the style of the Montana Ore Purchasing Company, and Mr. Heinze passed to its official head and directed its operations. Old heads looked askance at Mr. Heinze's temerity in venturing upon apparently so hopeless a course, but before the smelter was completed he had leased a mine and had begun operations therein. Powerful influences were brought to bear in the hope of eliminating Mr. Heinze from the field as a competitor for smelter business, but his young genius was not easily balked. Obstacles, seemingly overpowering, were met by bolder enterprises until he soon had acquired sufficient mining property to keep his smelter running to its full capacity independent of business from other sources.

Slowly it dawned upon the mining magnates of the camp that the Montana Ore Purchasing Company was a permanent factor in the field, and that Mr. Heinze was a Richmond who had come to stay. Mr. Heinze has commanded the admiration of a host of the citizens of not only the city of Butte, but of the whole State, by his nerve and daring, and, though many years behind the pioneers in his advent upon the scene, they see in him the kind of material of which the hardiest early settlers were made.

Mr. Heinze, as before stated, is a collegiate graduate, is cultured and refined, and has in him the promise of becoming one of Montana's great benefactors. He is unmarried and spends the greater portion of his time in Butte looking after the affairs of the company, living modestly and without the ostentation common to many upon whom fortune has so lavishly rained success.

A few years ago Mr. Heinze embarked quite extensively upon operations in the Rossland district of British Columbia, but the antagonism of the Canadian Government, through the instrumentality of subsidized corporations of that country, made his efforts extremely hazardous and he retired from the field, but not, however, before he had enhanced his wealth to a most satisfactory extent.

Mr. Heinze's wealth is not accurately known, but, as the largest owner of Montana Ore Purchasing Company stock, he unquestionably is many times a millionaire, and in his meteor-like elevation lies but another demonstration of what unwavering will and pluck will work out. Like his successful predecessors and contemporaries, Mr. Heinze has a large place in his heart for his old associates, and in this respect reflects the customs of his adopted State quite as happily as ever have those whose names precede this sketch.

COLUMBIA GARDENS.

For the past year a pleasure resort of the very first class has been accessible to the people of Butte. Across the valley and three miles east of the city one of the numerous cañons common to the Rockies has been utilized for this purpose. It has been preserved almost as nature made it, with additions only of such character as would enhance its inviting rusticity.

As the waters from the springs and melting snows high up the mountain side start upon their downward course, they join with others coming from different directions, and long before the bed of the cañon is reached a delightful stream babbles along over pebbly bottoms and gurgles over an occasional rock into inviting pools below. Luxuriant foliage fringes the banks of these tributary streams long ere their junction in the more level sweep below, and, as they emerge into one, a perfect Eden of green is massed about them, hiding, from a distance, the winding stream completely from the eye. Closer approach, however, but enhances the picture, and, as detail is added to detail, the withered soul creeps slowly out of its musty cave and breathes anew the joys of childhood. Willows and alders, with here and there a lonely pine, strayed from

BIRD'S-EYE VIEW, COLUMBIA GARDENS.

RANDOM VIEWS IN COLUMBIA GARDENS.
Pavilion in center.

its mountain side, have entwined themselves into inviting bowers and cozy nooks. Here nature has been aided for the comfort of man by the supplementing of rustic seats, the creation of shaded parks, by the clearing away of undergrowth, and the building of queer little bridges and aimless paths, while fountains for swans and goldfish add to the enchantment of this quiet retreat. Proceeding further, the trees terminate abruptly, and in all directions spread away before the view every conceivable device which may delight the heart, the mind or the soul of both old and young. Great rows of swings and merry-go-rounds attract the little ones like flies to molasses, while the more sedate mind has had every wish anticipated.

A great pavilion occupies a commanding position in the center of the grounds. Within its walls are café, banquet-room, smoking-room, refreshment booths of all kinds, and a dance floor of gigantic proportions and of ethereal surface, with its balconies for guests and orchestra, while surrounding the whole structure are broad promenade verandas and open-air band-stand.

The landscape gardener has given his touch of completeness to the scene by the creating of beautiful flower beds. Designs of striking likenesses have been worked out inside of odd-shaped plots created by the broad paths which wind about through the grounds. Still further toward the mouth of the cañon irregular paths lead through rustic ways, dotted by little pagodas of oriental style, to a delightful body of water, whereon glide lazily many boats at the will of idle pleasure-seekers.

The resort is peculiarly charming by reason of the fact that the citizens of Butte are deprived of a close communion with nature, due to the antagonism of the mineral nature of the soil to vegetation, and who, but for this beautiful retreat, would be denied the hallowing influences so necessary to the softening of natures and the expansion of the souls of men. Not only the thousands belonging to the laboring classes, but those of high estate have been quick to accept the privilege presented them to enjoy the pleasures of this bountifully endowed mountain retreat, so gratuitously thrown open to them, at no further cost than car fare, and immense picnic parties composed of people of both high and low degree are becoming daily sights within the grounds. Many excursions from about the State are scheduled for the coming year.

The gardens are under the control of the City Railway Company, but to Hon. W. A. Clark is entitled the honor for having provided so necessary a public institution, that gentleman, as president of the system, having been its instigator and enthusiastic patron. Something like $50,000 has been spent within the year in adding to the general improvement of the grounds. Electric light, fire, sewerage and water systems have been extended about the entire forty acres comprising the gardens, the first-mentioned system rendering the grounds as attractive by night as by day. In addition to these improvements a fine botanical garden and zoölogical collection are planned for the near future to supplement those already started, and which will be free to the public. Over 80,000 plants have been taken from the hothouses operated upon the grounds and have been placed in the numerous beds, and many rare trees comprise those which line the walks and paths.

SCHOOL OF MINES.

The Montana State School of Mines is located on an eminence just west of the city limits of Butte. The building is constructed of brick and stone in the Renaissance style of architecture. It is practically fireproof and has been considered the finest public structure in the State. It was erected in 1896-98, and has recently been supplied with $15,000 worth of furniture and apparatus. Everything that goes to make up this furnishing is of the best quality and of modern construction. The illustrations here given are all half-tones taken from photographs and will give a correct idea of the style of the building and its equipment.

The institution was opened for the reception of pupils on September 11, 1900, and at that time freshman and sophomore classes were formed. The courses of study adopted require four years for their completion and lead either to the degree of Mining Engineer or Electrical Engineer, according to the lines of topics chosen by the student. The requirements for admission and

INVITING BOWERS AND COZY NOOKS, COLUMBIA GARDENS.

for graduation correspond closely to those of similar institutions in other States.

This school is under the control of the State Board of Education and a local Board of Trustees. The members of the local board are: John E. Rickards, ex-governor of the State; James W. Forbis, W. Y. Pemberton, ex-chief justice; George E. Moulthrop and Joseph V. Long. The members of the faculty are: Nathan R. Leonard, president and professor of mathematics; William G. King, professor of chemistry and metallurgy; Alexander N. Winchell, profes-

SCHOOL OF MINES.

Lecture Room of Laboratory. Lecture Room, Physics Department.

CLASS ROOMS, SCHOOL OF MINES.

Laboratory, Working Room. Drawing Room.

ENTRANCE AND FOYER, SCHOOL OF MINES.

sor of geology, mining and mineralogy, and Charles H. Bowman, professor of mechanics and mining engineering.

The Act of Congress providing for the organization of the State granted 100,000 acres of land for the establishment of a School of Mines. The legislature of the State located the school at Butte, and has made liberal appropriations for its equipment and current expenses.

The large mineral resources of Montana and the vast amount of capital employed in their development have made mining the chief industry of the State. The School of Mines is therefore an object of the greatest interest to the people of this commonwealth. It is hoped that in a comparatively short time this institution will be in possession of very large and valuable collections of minerals, mine models, etc., illustrating the resources of the State and the latest and best methods in the extraction and treatment of ores. The City of Butte is the greatest mining center in America. The thousands of trained superintendents, engineers, metallurgists and practical miners here employed constitute an environment that will prove of inestimable value to the School of Mines.

THE PAUL CLARK HOME.

"To help the worthy poor to help themselves" was the motto adopted by a little band of charitable women one fall day in 1897. About three years later, or upon Friday evening, November 16, 1900, there was formally opened, by an unqualifiedly successful charity ball, the Paul Clark Home as a fitting monument to their untiring labors, and but for which this beautiful structure would, perhaps, never have been erected.

So generously did their motto and the high standing of the organizers appeal to public sentiment that success followed success rapidly until their organization — the Associated Charities — soon became the recognized leader in charitable work in the city. The charter members of the infant organization numbered about fifty, but, so enthusiastic have been their efforts, that in the short years since its incorporation the membership has increased to nearly 200 — exclusively of ladies — most of whom are leaders in social and religious life in the city. The first officers and trustees of the organization are entitled to distinct mention, together with all honor that might go with it, for upon the proper shaping of the association's affairs at its inception rests, in large measure, the credit for subsequent successes. Their names are: President, Mrs. J. M. White; first vice-president, Mrs. John Noyes; second vice-president, Mrs. A. S. Christie; secretary, Mrs. Irene Morshead; treasurer, Mrs. A. M. Wethey; trustees, Mrs. J. M. White, Mrs. C. W. Clark, Mrs. Annie E. Hammond, Mrs. Jennie H. Moore, Mrs. Sarah Broughton, Mrs. Ruth Burton and Mrs. John Noyes. Aside from its membership, nearly every business and professional man of note in the city is numbered among the association's list of donors.

PAUL CLARK HOME.

The objects of the association are: To help the worthy poor to help themselves, to visit and assist the poor, relieve their distress by providing physicians, nurses, food, clothing, fuel and whatever may be necessary in their particular cases.

A home was secured where those seeking employment and without means could remain temporarily, the end sought in all cases being to help the assisted to help themselves. In the language of its noble president, Mrs. J. M. White: "To put one family beyond the necessity of charity is more useful than to tide twenty over into next week's misery."

Truths are often best left unsaid, but it is felt to be the fact, concurred in by every one and whose relation is a pleasure to all, that if a canvass were made for an explanation of the large measure of success which has attended the association's history, the unanimous reply would be "Mrs. White." While a score of others have loaned their every thought to the upbuilding of the association and sacrificed their personal comforts in ministering to the wants of others, yet undoubtedly is due to Mrs. White the credit for the growing success of this ennobling work. Hers is the genius, the tact, the farsightedness, the generalship, which, combined with a sweet, pure woman's heart, has tided the association over the dark places and brought to its support assistance which otherwise might not have been enjoyed.

Mrs. John Noyes has always been a most able lieutenant, making light the many burdens and lending her best efforts at all times, and not infrequently of her large means, for the success of the association.

Another earnest worker, as well as donor, is Mrs. C. W. Clark, who, as one of its first trustees, has been identified with the work from its start and has ever been willing with her means and kindly personal effort to further the best interests of the association.

Scores of others might be mentioned who have loaned both effort, thought and means to the success of this God-given task, whom many a hungry, dejected soul has learned to bless from a touched and softened heart.

As the association grew in age and stature, its needs rapidly multiplied and the small frame building originally occupied by it proved wholly inadequate to meet them. The burning question of how to meet this new demand with the limited funds at its command lay heavily upon the hearts of more than one zealous member for many days before its solution was reached.

Meanwhile, Mrs. White, anticipating events, had taken the matter up quite seriously with Hon. W. A. Clark, suggesting to him the necessity for an ample home for the association, wherein its plans might be worked out unhampered. The suggestion, seemingly, took root in Mr. Clark's mind, and some months later, while in the East, he notified Mrs. White to confer with architects in the drafting of plans for a building to cost some $20,000. His communication also stated that his immediate family would supply the furnishings for the institution in all details. This was supposed to represent an additional outlay of $10,000.

SUN PARLOR

Plans were drawn and work begun, and, as construction and furnishing progressed, the original sums appropriated became exhausted many times and, as often as this occurred, new appropriations were generously made until the building completed represents a total cost of not less than $50,000. The only condition imposed upon the association was that the name of the institution should be the Paul Clark Home, in memory of Mr. Clark's youngest son, who, during his lamentably short life, devoted so much to deeds of charity.

Mrs. White, at the request of Mr. Clark, assumed complete control of the plans, material, furnishings, etc., and, so great a burden did this impose upon her, she was forced to withdraw from active charge of the Associated Charities, and, for the last two years, has given her entire thought and time to her new duties.

The completed structure, both exterior and interior, is a most beautiful adornment to the city and a lasting monument to Mr. Clark's public-spiritedness. As it shall become better known, its usefulness will impress more deeply the citizens of Butte and a still keener appreciation of it is in store. That it will immeasurably facilitate the work of the Associated Charities is palpable, and it is predicted that untold suffering and distress will be relieved by its agency.

The building is constructed of rustic brick with granite facings and arched entrance, and finished inside with highly polished oak and artistic red brick fireplaces, and all the latest devices for perfect ventilation. Its furnishings are in perfect accord with the purposes for which they are to be used, and, as shown by accompanying illustrations, are simple but elegant.

READING-ROOM.

The main building is three stories in height, with a large, well-lighted basement. In the rear is a hospital building one story high and divided into two large wards. Connecting the main building and the hospital is a large sun parlor, the sides of which are composed almost wholly of glass. The illustration but faintly shows what a great boon this feature will prove to the convalescing patient.

The first floor of the main building is comprised of office, reception, writing, reading and dining rooms, day nursery, wardrobe, soup-room, pantries and kitchen. On the second floor are the sleeping-room for the day nursery, seven single and three double bedrooms, matron's room, two large linen-closets and bath and toilet rooms, fitted up with every convenience. The third floor has six single and two double bedrooms and a trunk-room. Each room has its convenient closets. The basement contains a large laundry, drying-room, four large rooms for industrial purposes, bath, toilet-rooms, wash-rooms, fumigating closets, store-rooms, furnace-room and large fuel compartments.

While the Paul Clark Home was originally built for the purpose of serving as an auxiliary to the Associated Charities in the prosecution of a larger work, the idea has more recently been advanced with considerable reason that it presented untold opportunities complete in themselves. Many of those most actively interested in its future feel that a great economic benefit may be worked by converting the institution into an industrial home. In so doing opportunity will be given many worthy persons for fitting themselves to become independent, thereby carrying out on a larger scale the idea of "helping the worthy poor to help themselves." Even this course, however, would not deprive the Associated Charities of the benefits originally sought in the erection of the Home, the only result gained being a broadening of the institution's possibilities.

A feature which will be adopted in any case will serve as a great blessing to many dependent women of large families, who at present are restrained from earning the means necessary for their support. A nursery has been provided for the care of the children of these women who are of too tender an age to be left at their homes, and by this means their mothers will be enabled to seek employment otherwise denied them, thereby being relieved of "next week's misery." Even within the institution, if present plans are carried out, opportunity will be granted a limited number of these mothers to gain a livelihood by performing work from which the institution may derive a revenue, and, so far as possible, the children will be given an opportunity to learn vocations whereby they may make themselves self-supporting. It is a grand work and both the donors of the institution and those most closely interested in its direction may be sure of a full measure of reward for their liberality and Christian charity.

MINING.

The widespread growth encountered throughout every other portion of Butte is multiplied by a much larger number with respect to the mines. True, the old Travona district, the first active scene of quartz-mining, to the southwest of the city, is deserted. An occasional weed-grown shack sheds the elements for some wayfarer, but mining is dead, and the old surface structures, fast crumbling to decay, are uncanny in the memories they conjure in the mind. Missoula Gulch, though fast losing its identity northward under the hand of new improvements, here — in the same uncouth desolation in which the last pick left it — eloquently pleads for reverence.

Walkerville, to the extreme north, too, has its mournful tale. There, gaunt in their deserted grandeur, stand the gigantic mine and smelter structures of the Lexington, Moulton and Alice, the two former never to resume their lives of usefulness, and the latter probably doomed to the same destiny. The Magna Charta, Valdemere, and other once important, though less noted, mines, also, in their crumbling state, but add to the truth that mining in this section is fast

"Travona," the first paying silver mine.

Stack of Centennial Smelter.

"Star West."

CRUMBLING RELICS OF THE TRAVONA DISTRICT.

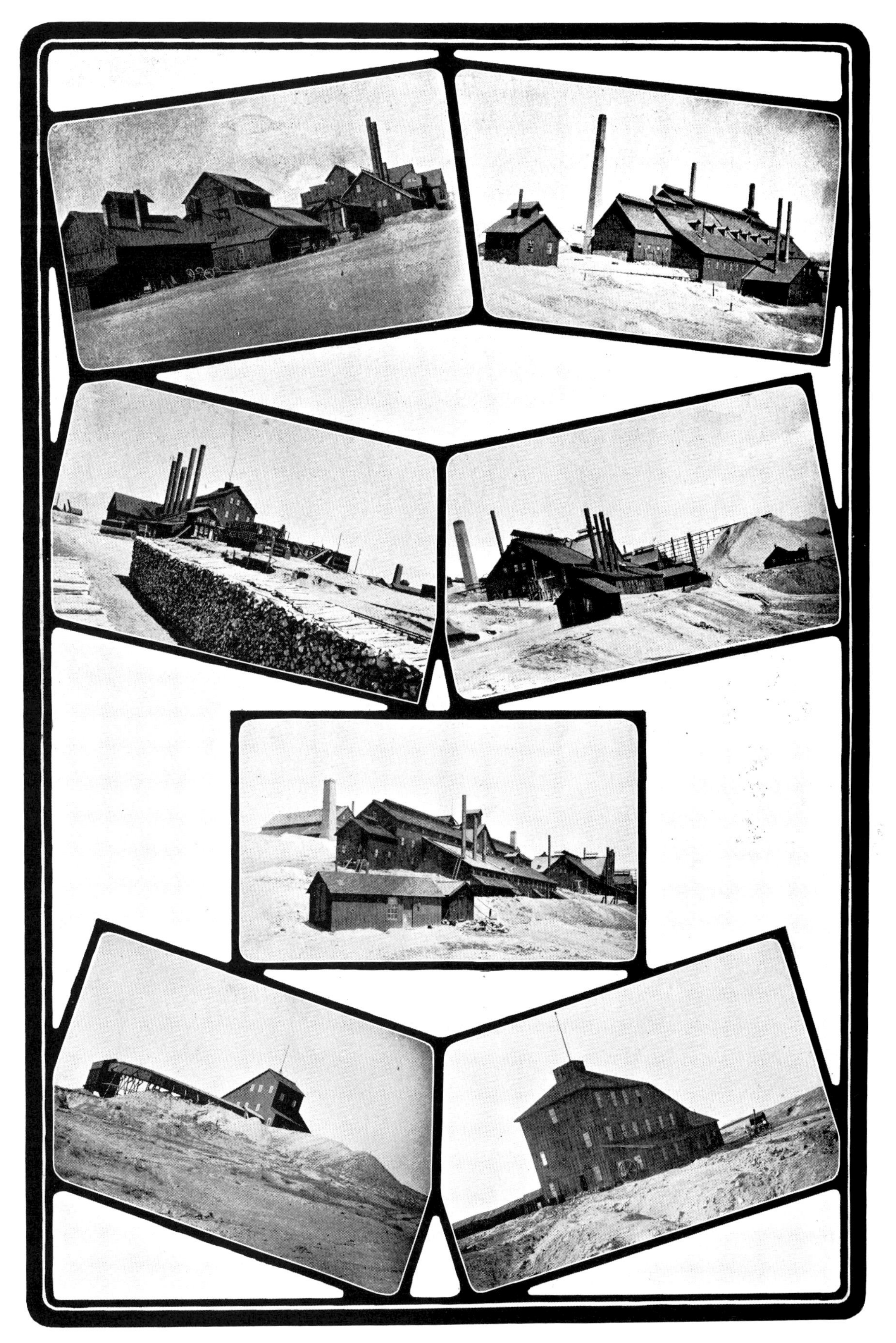

Lexington Mine.
Alice Mine.

Moulton Mine and Smelter.

Lexington Smelter.
Alice Smelter.

becoming a lost industry. Some three miles west from the city's limits, the Blue Bird and other noted silver properties, contemporaries and lusty rivals of the larger ones in Walkerville, have yielded up the ghost and are being demolished.

But here ends the list of decline and decay. Overwhelmingly outbalancing it, is the list of increase and growth in mining activity elsewhere. "The hill" is the old hill still, but a greater mystery. Hundreds of claims, under the control of a few owners, cover every available inch of this wonderful spot from a point immediately southeast of Walkerville to Meaderville. But the list does not stop here. Not a square inch of ground is there within the city's limits and running beyond, to east, south and west, but what is a prospective mining claim, titles to which are almost universally exempted in all deeds of conveyance transferring surface rights. In other words, the instance is rare indeed where a transfer of surface rights by sale does not exclude the mineral deposits beneath such surface, so impressed is the whole community with the idea that copper deposits on "the hill" do not end at its borders, but extend downward on either side of the same, and into the respective valleys. So well grounded is this idea that there are many who hold that, as the many leads are extended under the hill, they will take their owners completely under the city to the south or across the valley to the east and into the main range of the Rockies. Color is given this theory by discoveries, while excavating, of several most important leads in the very heart of the business and residence sections of the city, as well as in the lowest levels of the valley to the east. Even beyond and along the opposite side thereof and up the main pass through the divide, some five miles east, important operations are being prosecuted. The Homestake property is an important member of this group and promises new and important fields in entirely new quarters.

RICHEST MINING SPOT ON EARTH.
Famous "Anaconda Hill," showing different mines of the Parrot lode, the scene of early-day mining.

Not one of the dozens of mines which honeycomb "the hill" from either side has failed to retain its "lead" at whatever depth it has sunk its shaft and the general tendency of such lead is to widen as greater depth is reached. As previously stated, copper is the ore primarily sought in the whole Butte district, and the precious metals are but by-products. In some cases these latter furnish sufficient revenue to maintain the operations of the entire property and the copper becomes a net profit to its owners.

In all there are some 225 mines in the immediate environments of the city, though many, it is true, are but infantile in their proportions and their owners are only performing their legal "representation" work. In the neighborhood of 13,000 men are employed in these mines and smelters of this city. It may be added, in this connection, that two other cities — Anaconda and Great Falls — owe their existence to the smelters, which are owned and operated at those points by the mine owners of Butte. The largest institutions of which this country at least can boast of that character are situated in these three cities and give employment to thousands of men.

To enumerate in detail all of the numberless phases of the mining conditions existing throughout the district would be in turn an endless task and a tiresome repetition. Suffice to say that the district today is the most important, from a mining view, of any district on earth. For Montana's reputation as a great mining State, Butte is almost wholly responsible. In fact, the statement is often made that "Butte is Montana." Though increased mining activity is beginning to develop throughout other portions of the State and many old sections are holding their own, the gross output therefrom at present is insignificant compared with the Butte section.

The relation of this section to outside sections is shown in the following table of outputs for 1899:

	COPPER.	SILVER.	GOLD.	LEAD.	TOTAL.
Silver Bow	$40,882,492	$12,742,893	$1,292,447	....	$54,917,833
Outside	59,414	9,043,942	3,526,710	$909,410	13,539,476
Totals	$40,941,906	$21,786,835	$4,819,157	$909,410	$68,457,309

Too much importance can not be attached to the fact that in this year the output of the State increased just thirty-three per cent over the previous year.

With Butte's mineral preëminence in Montana established, its relation to the entire country and to the world as a copper-producer is worthy of

"THE HILL" FROM SOUTH SIDE.

consideration. Perhaps nothing that could be said upon the subject would speak more eloquently than a short excerpt from the annual report of Hon. E. B. Braden, United States Assayer in Charge for Montana, which reads as follows:

"Previous to 1882, 80 per cent of all the copper of the United States came from the mines bordering on Lake Superior. In the following year the Lake Superior region produced 51.6 per cent, Arizona 27 per cent, and Butte 21.4 per cent of the domestic copper. The percentage of the Butte output continued to increase steadily, and in 1887 it became greater than the yield from the Lake Superior district. This lead has ever since been advanced until, in 1898, when 60 per cent of all the world's copper was supplied by the United States, Butte furnished 41 per cent, Lake Superior 30 per cent, and Arizona 21 per cent of all the domestic production. Butte thus practically furnishes a quarter of the copper product of the world."

The ratio of growth in mineral output in Silver Bow district during the period of 1882-1899 is shown in the subjoined table from Mr. Braden's report:

Year.	Gold, Fine Ounces.	Silver, Fine Ounces.	Copper, Fine Pounds.
1882........	12,093,750	2,699,296	9,058,284
1883........	14,560,875	3,480,468	24,664,346
1884........	21,776,006	4,481,180	43,093,054
1885........	13,838,297	4,126,677	67,797,864
1886........	31,223,450	5,924,180	57,611,485
1887........	48,175,743	6,958,822	78,700,000
1888........	44,320,062	8,275,768	98,504,000
1889........	31,652,325	6,560,038	104,589,000
1890........	25,704,730	7,500,000	112,700,000
1891........	29,395,356	7,985,089	112,383,420
1892........	36,222,560	8,311,130	158,413,284
1893........	33,807,877	6,668,730	159,875,490
1894........	36,768,015	7,561,124	185,194,385
1895........	41,493,363	10,051,760	197,190,650
1896........	59,815,755	11,120,731	228,886,962
1897........	54,198,037	10,710,815	236,826,597
1898........	55,038,589	8,996,555	216,648,077
1899........	62,038,377	9,855,831	245,245,908
Total.......	652,368,167	131,268,203	2,337,382,824

To the above may be added the fact that the world's output for the following year, or 1899, showed a slight increase over the preceding one, so divided as to maintain the percentage deduced by Mr. Braden.

Figuring copper at the price prevailing during the year of its production, the revenue from this commodity represents a gross sum of $284,531,-746. If the same price had been enjoyed during these years, as will doubtless maintain, if not increase, in the future, the copper output to date would have represented a gross revenue of about $400,000,000.

It is doubtful if the people of this, or any other, section fully comprehend the importance of Montana as a mining State. It has been more generally classed as one of the States of the "mining West," many other States enjoying the same general reputation that should specifically apply to Montana first — placing even the much boasted Colorado mineral wealth well into second position. From the table shown on page 6 it is

THE "SMOKEHOUSE."

Discovered within the year in heart of city, while excavating, and sold for half a million.

believed that the true relation of Montana to the mining industry, not only of the West, but to the whole country, will be universally recognized for the first time. In this table it is impossible to show the precise value of iron production. This kind of ore is treated or reduced to pig iron at points foreign to the mine, and no credits are given to the producing section, values being placed upon the pig iron after treatment of the ore. Of the three sections producing iron ore, the Lake Superior region produces approximately three-fourths, the Southern States two-thirds of the remainder, and all other States but one-third. As the Lake Superior region undoubtedly is synonymous with Upper Michigan as regards the iron industry, for purposes of calculation, Michigan is credited with three-fourths of the total value of pig iron production in the table. This, no doubt, is greatly in excess of the true value of the crude ore, but hardly sufficient to change the relative positions of the States as named.

LOOKING EAST FROM MEADERVILLE.
Showing extension of operations across valley toward Rockies.

It will be seen that Montana is easily the third wealthiest in point of production of all the mining States of the country, Michigan leading, with Pennsylvania second. In copper production Montana leads all other States, approximately 40 per cent of the nation's output coming from Butte. First place in silver production also belongs to Montana, the greater percentage of which comes from Butte as a by-product in copper mining. Gold, coal and lead make a most creditable showing, especially in the two latter, considering the brevity of operations in those fields.

With the prestige thus enjoyed and with the prosecution of extensive development work all along the line throughout the whole State, the prediction seems quite permissible that Montana will not only continue to hold her own, but will forge ahead each succeeding year, until she not only leads the entire West, but will be a competitor for first place as the greatest mining State of the Union.

As for Butte's part in the State's great future, precise prediction would appear presumptive. That it will continue its present tremendous lead no competent authority doubts. That it will enjoy the steady growth of the past, trebled and quadrupled by virtue of increased mining activity, aided no little by a logical growth along commercial and manufacturing lines, seems modest enough to predict; but, without infallibility, a prediction less optimistic would seem absurd, and every sign but reinforces its truth.

A more detailed reference to the principal mines of the city, located, without exception, upon or contiguous to "the hill," together with smelters operated in connection therewith, lying along the valley to the south, follows, and verifies eloquently many statements preceding, which, but for such corroboration, may have been open to the charge of too much zeal.

It has been shown elsewhere that the large majority of the mines and smelters of the Butte district are controlled by a few large mining corporations. These companies control in overwhelming proportions all of the mineral rights underlying "the hill" on either side from Walkerville to Meaderville and, in most cases, the surface rights as well.

The corporations thus referred to are the Anaconda Copper Mining Company, Colusa Parrot Mining Company (Clark interests), Boston and Montana Mining Company, Butte and Boston Mining Company, Montana Ore Purchasing

Company, Colorado Mining and Smelting Company, Parrot Mining Company and the Largey estate interests.

Smelters are operated in Butte in connection with the Colusa Parrot, Butte and Boston, Montana Ore Purchasing, Colorado Mining and Smelting and Parrot Mining Companies' properties, while the Anaconda mines send their ores to their smelters at Anaconda over their own railroad — the Butte, Anaconda & Pacific — and those of the Boston and Montana are sent to their Great Falls works over the Montana Central Railroad for reduction.

In addition to these corporations there are any number of smaller ones and, in many cases, of individual owners whose properties are scattered throughout the same area covered by the larger corporations and which are developing into first-class propositions. The ores from these mines are sent to the various smelters above enumerated for reduction.

MOOSE.
B. & M. Co's property. Crest of Hill, east of Walkerville. Depth, 300 feet. Employs 30 men.

In making specific reference to the mines of Butte by means of illustration, the method thus employed of reinforcing Butte's claims to first position among the mining districts of the world presented such an endless task, so characterized by a seemingly tireless repetition, that the necessity of confining the list to the larger interests appeared mandatory. If taken through to a finality a work of twice the thickness would be required.

All through the area most generously endowed are hundreds of shafts, marked by gallows frames and dumps of various sizes, which mark the varying progress of different claims, all of such an identical appearance that no distinctive feature could be shown, save in the name. This is true of many sections of the city itself, and the instance is not rare of mining operations on a limited scale being conducted in vacant lots lying between two dwellings, stores, etc.

The order observed in the following illustrations, it is thought, will give a clearer idea of the general mining situation in connection with the distribution of the mines according to district than if classification were made under heads of the various corporations. The idea has been to pick up the thread of the earlier portion of this chapter, and by illustration show where active mining operations begin as compared to the decay in other sections. It will thus be seen what is meant by the overwhelming increase of the new over the old. The list begins at Walkerville, proceeds thence to Centerville along the southern and western slopes of "the hill" as it zigzags in its southeasterly direction until Meaderville is reached, and thence north along the eastern slope until operations practically cease.

PAULIN.

Washoe property. Extreme southwestern slope of Hill. Centerville district. Operations suspended pending completion of Washoe smelters at Anaconda. Employed 100 men. Depth, 1,200 feet.

BUFFALO.

Anaconda property. Southwest slope of Hill. Centerville district. Depth, 1,600 feet. Employs 100 men. Weekly output, 500 tons.

LITTLE MINAH.

Parrot property. South slope of Hill. Centerville district. Depth, 700 feet. Employs 35 men.

MOUNTAIN CONSOLIDATED. NO. 2.

Anaconda property. South slope of Hill. Centerville district. Depth, 1,800 feet. Employs 300 men. Weekly output, 2,500 tons.

MOUNTAIN CONSOLIDATED, No. 1.

Anaconda property. South slope of Hill. Centerville district. Depth, 2,000 feet. Employs 550 men. Weekly output, 4,000 tons. Has 17 exits.

WEST GRAY ROCK.

B. & B. property. South side of Hill. Centerville district. Depth, 700 feet. Employs 50 men.

CORA.

Under lease to M. O. P. Co. Centerville district. Depth, 400 feet. Employs 50 men.

EAST GRAY ROCK.

B. & B. property. Crest of Hill. Centerville district. Depth, 1,800 feet. Employs 150 men.

GREEN MOUNTAIN.

Anaconda property. Southwest slope of Hill. Centerville district. Depth, 2,200 feet. Employs 360 men. Weekly output, 2,900 tons.

DIAMOND.

Anaconda property. Crest of Hill. Centerville district. Depth, 2,200 feet. Employs 550 men. Weekly output, 6,000 tons.

BELL.

Anaconda property. Crest of Hill. Centerville district. Depth, 1,800 feet. Employs 275 men. Weekly output, 4,000 tons.

PARNELL.

M. O. P. Co's property. Northwest slope of Anaconda Hill. Depth, 700 feet. Employs 35 men.

ANACONDA.

On "Anaconda Hill." Largest mine in city. Scene of Mr. Daly's early activities and nucleus of all Anaconda properties. Depth, 1,800 feet. Employs 1,400 men. Weekly output, 9,000 tons.

NEVERSWEAT.

So named because men do not sweat below ground. Southwest of Anaconda mine. A. C. M. Co. property. Depth, 2,000 feet. Employs 600 men. Weekly output, 4,500 tons.

RAMSDELL'S PARROT.

South of Neversweat. Named for Joe Ramsdell. A. C. M. Co. property. Depth, 600 feet. Employs 200 men. Weekly output, 1,600 tons.

COLUSA PARROT.

West of Ramsdell Parrot. Clark's property. Depth, 1,600 feet. Employs 350 men.

PARROT.

Before destruction by fire the past year. Depth, 1,300 feet. Employed 350 men. West of Colusa Parrot.

AFTER FIRE.

Ruins were immediately removed and new works in course of erection.

STEWART.

North of Parrot mine. Clark property. Depth, 1,100 feet. Employs 200 men.

NIPPER.

Northeast of Stewart. M. O. P. Co. property. A new shaft being sunk. Depth, 800 feet. Employs 150 men.

ORIGINAL

On lode where was found a hole dug presumably by Indians. South and west of Hill. Clark property. Depth, 1,300 feet. Employs 200 men.

GAGNON.

Adjoining Original on west. Col. M. & S. Co. property. Depth, 1,800 feet. Employs 300 men. Anaconda Hill in background to the east.

BLUE JAY.

B. & B. property. South side Anaconda Hill. Depth, 1,250 feet. Employs 300 men.

MOONLIGHT.
Washoe property. South side Anaconda Hill. Depth, 1,500 feet. Employs 350 men.

ST. LAWRENCE.
Immediately south of Anaconda mine. First acquisition of Anaconda Company after purchase of Anaconda. Been on fire for ten years. Depth, 1,600 feet. Employs 1,100 men. Weekly output, 2,500 tons.

PENNSYLVANIA.
B. & M. property. South side Anaconda Hill. Depth, 1,430 feet. Employs 300 men.

SILVER BOW, NO. 1.
B. & B. property. Southeast slope of Hill. East Butte district. Depth, 1,000 feet. Employs 185 men. Horses are stabled on 400 level and never come to surface.

SILVER BOW, NO. 3.

B. & B. property. South of Hill. East portion of city. Depth, 500 feet. Employs 50 men.

MOUNTAIN VIEW.

B. & M. property. Highest point on Anaconda Hill, facing Meaderville. Depth, 1,750 feet. Employs 125 men. Has 14 exits.

RARUS.

M. O. P. Co's property. First mine acquired by F. Aug. Heinze. East slope of Hill. Meaderville district. Employs 250 men. Depth, 1,100 feet.

BERKELEY.

B. & B. property. East slope of Hill. Meaderville district. Depth, 700 feet. Employs 70 men.

MINNIE HEALEY.

Title in litigation. Base of east slope of Hill. Meaderville district. Depth, 1,800 feet. Employs 150 men.

LEONARD.

B. & M. property. Base of east slope of Hill. Meaderville district. Depth, 1,130 feet. Employs 180 men.

WEST COLUSA.
B. & M. property. East side of Hill. Meaderville District. Depth, 1,370 feet. Employs 175 men.

EAST COLUSA.
B. & M. property. Base eastern slope of Hill. Meaderville district. Depth, 800 feet. Employs 65 men.

HIGH ORE.

Anaconda property. East slope of Hill. Meaderville district. Depth, 2,200 feet. Employs 300 men. Weekly output, 2,300 tons.

MODOC.

Anaconda property. East slope of Hill. Meaderville district. Depth, 1,000 feet. Employs 100 men. Weekly output, 1,200 tons.

SPECULATOR.

Largey estate property. East slope of Hill. Farthest north in Meaderville district. One of the best equipped mines in city. Depth, 1,200 feet. Employs 70 men.

MONTANA ORE PURCHASING COMPANY'S SMELTER.

East slope of Hill. Meaderville district. Scene of Heinze's initial operations. Employs 350 men.

PARROT SMELTER.

Closed at present, due to fire in mine. Extreme southeastern portion of city. Employed 250 men.

BUTTE AND BOSTON SMELTER.

Southeast slope of Hill. Meaderville district. Employs 350 men.

COLORADO SMELTER.

Col. M. & S. Co. property. Extreme southwestern limits of city. Employs 300 men.

BUTTE REDUCTION WORKS.

Extreme southern portion of city. Clark property. Employs 350 men.

THE RED METAL.

⁂

STEEL STACK 112 FEET HIGH.

THE mechanical steps, as followed in a general way throughout the various developments of a mine, from its inception to a stage which surpasses the imagination and makes description impossible, can not fail of a lively interest.

It has already been stated that the mineral right underlying every available inch of surface ground for a large area is firmly held by virtue of location, purchase or otherwise, oftentimes by parties other than the surface owners. That these rights are subject to purchase is as true, in a majority of cases, as is the fact that the surface right may be bought for sufficient consideration.

Whoever is fortunate enough to own the mineral right of a particular portion of this

THE PRIMITIVE WINDLASS.

wonderful area has made an important start. Failing this, he is compelled either to purchase such right or to "lease and bond" the same. By the terms of the latter agreement the lessee undertakes to develop such claim to a certain depth, to employ a given number of men, to timber all shafts and drifts, and to pay the lessee a royalty on all ores extracted from the claim. The lessor, on his part, agrees and undertakes to sell such claim to the lessee within the time stipulated in the lease at a price agreed upon therein. Thus the lessee becomes the nominal owner for the time being, with the right to purchase. By this process he either takes up the work where the owner or a previous lessee has

HAND ORE CAR.

laid it down, or starts at the very beginning and breaks ground for the first time upon the claim. In the latter case a "prospect hole" is the first step taken and the thing sought is palpably a "lead." If the lead is found and a vein of sufficient value is uncovered to warrant it work is continued.

As the vein becomes more clearly defined and the prospect hole deepens, excavation by the usual methods employed in digging a hole or ditch becomes impracticable, and the first surface structure, a "windlass," is erected over the opening or the mouth of the hole, and the latter is thenceforth designated as a "shaft." At this juncture, to prevent shifting of the earth from the sides of the shaft, heavy timbers are installed to brace the side walls, and in this manner the shaft is boarded in on all sides as it deepens. A large bucket is lowered from or raised to the surface by means of the windlass, and as the excavated matter is hauled to the surface it is "dumped" into a primitive ore bin if it carries ore values, otherwise onto the waste dump and there to remain.

It is the practice of all mines in this section to establish a "level" at every hundred feet of the shaft, and when this important stage is reached the mine takes on a new significance. While work continues to be prosecuted at the "sump" end or bottom of the shaft to secure greater depth, a "drift" or tunnel is also started at a right angle to or in a horizontal direction from the shaft and approximately in the direction of the ore body. Thus is established the first or "one hundred" level. Tracks are now laid along the floor of the drift, and along these "ore cars" are run from the point of operations to the "station" at the shaft end of the drift.

Meanwhile the surface structure has taken on a new aspect. The windlass has outlived its usefulness, as greater hoisting power is required, and in its place a horse "whim" is erected. This is soon followed by a more substantial though diminutive hoisting device known as the "gallows frame," made of heavy timbers, and steam power is substituted for horse. In many instances this order of things is not found expedient, and the gallows frame is erected at the very outset. The bucket, too, has gone the way of the windlass, and the whim and a steel skeleton-like contrivance known as the "cage" is introduced as the vehicle for hoisting purposes, the ore cars being run upon them and car and cage raised to the surface. A more pretentious ore bin, too, has been constructed and tracks are run both to it and to the waste pile or dump. To one of these the cars are taken and unloaded, propelled by either man or horse power.

As the drift extends new leads are encountered and new drifts or "crosscuts" are established. By the time this stage of development has been reached, if not previously found necessary, although it ordinarily has been from the very outset, the rebellious nature of the ground has compelled the use of other implements than pick and shovel, and resort is made to giant explosives in breaking the way. Before these explosives can be used, deep holes are first necessary and, for the purpose of making them, a drill team of two men is put to work, one of the team holding the drill — a cold chisel sort of device — while the other wields his sledge, pounding the drill deeper and deeper into the face of the rock.

HORSE WHIM.

No more machine-like exhibition can be found anywhere than is displayed by this class of miner. Every stroke is in perfect rhythm, the sledge striking the drill squarely upon the head with unerring precision. Whether swung from above, below or either side, the result is always the same, the sledge always finding the mark. As the first drill, by reason of its inferior length, renders its holding dangerous to the hand, it is neatly extricated, while yet the sledge is in motion for the succeeding stroke, and another inserted of greater length, and the blow falls as before, squarely upon the head of the new drill and not a motion or a second wasted. Likewise with the changing of positions of the two men. As the sledge descends upon the drill and begins to describe a new circle for another blow the member of the team who previously held the drill grasps the sledge without so much as a pause in its course, the other member grasps the drill and the blow descends as before, squarely upon the head of the drill and not a motion or a second wasted.

Thus turn about, first one and then the other wields the sledge until the hole is complete. Another hole and still another follows, until a half dozen or more have been drilled into the face of the rock. Then is inserted the explosive, a fuse is attached and all withdraw to safe distances until the report tells them to return. Thus foot by foot and yard by yard the drift is pushed further and further along toward or through the ore body, cars speedily removing the debris to the station and thence to the surface, while new tracks constantly keep apace with the progress of the drift.

"TWO-DECKER." TWO-POST GALLOWS FRAME.

Meanwhile the ore body is attacked along the ceiling of the main drift by the "stoping" process, great quantities of ore being blasted away from the top of the drift and, falling to the floor thereof, are carried to the surface. If the character of the ground through which the drift is run is of a yielding nature common to earth formation, as the drift progresses its sides are timbered by strong beams and boards, and, as the stoping process continues, a timbered roof is built to the drift, which becomes the first floor of the stope. In many cases the character of the ground worked is of such a nature that very little or no timbering is required, the safety of the mine or miners in nowise being jeopardized by the absence of the same.

Skeleton frames of powerful scaffolding are erected, however, from the floor of the drift, even in this class of ground, in order to permit of the laying of additional floors in the stope above. Stoping operations are exclusively followed, in working from one level to another, larger quantities of earth being more easily dislodged by working from the ceiling, and the drift floors are

CAGE.

thus kept intact for the rapid transit of cars to and from the station. In many mines of Butte these stoping chambers are three hundred or more feet wide, but the presence of so much timbering renders their illustration impossible. Along the first few levels, which follow in regular order of one hundred feet as the shaft descends, the stoping process does not constitute so large a portion of the operations, the vein at this depth not having developed sufficient width nor value to warrant it. As greater depths are reached, however, the mine expands in every direction.

CARRYING ORE TO THE BIN.

Then comes the widening of the shaft. From a "one compartment" shaft, admitting of the passage of but one cage, it graduates into a two-compartment shaft and another cage is introduced, each operated separately from the other. A third compartment is also found necessary in most cases, for the running of all manner of pipes and wires necessary to the successful operation of the mines. Unlike many other kinds of mining, no gases are encountered in the mines of Butte. Notwithstanding this fact, every care is taken to supply all levels with a constantly changing air, thus overcoming in short order the bad effects caused by blasting, etc. Immense engines for this purpose are installed at a convenient point near the mouth of the shaft, and are most carefully inspected. Water, too, is an element to be figured upon no little in the operation of the mine, large quantities of it seeping through the sides and roofs of the respective drifts, and upon its immediate elimination from the workings very much depends.

Compressed air for the operation of the air-drills throughout the mines is also found necessary with the introduction of this class of drill in substitution for the hand drill. These drills are many times more efficacious than the more primitive drill heretofore described, operating much the same as an auger, boring their way into solid rock with the ease that that instrument would into a pine board. Power to operate them is furnished in compressed air, and when it is realized that no less than 250 of these machines are being operated in one single mine, the Anaconda, this feature alone takes on a gigantic significance.

HAND DRILLING.
1,200-foot level, Parrot Mine.

Aside from the proverbial candle with which all miners go armed, a complete system of electric lighting is installed in every mine, lights running along the roof of all drifts and crosscuts, lighting the dark interiors and permitting of the most expeditious prosecution of all branches of operations. The lamp is an unknown article below ground in this section.

For these pipes of all kinds, wires, speaking tubes, etc., the creation of the third compartment in nearly all mines is a most necessary essential.

But the expansion does not stop here. Again has the surface appearance undergone a change. Still again has the gallows frame proved its insufficiency. A greater hoisting power has been found necessary by reason of increased quantities of ore and waste resulting from the extension of shaft, drift, stope and crosscut, and with this necessity is born the steel gallows frame. So popular is this style of frame becoming that many of the mines are tearing down their old shafthouses, which enclosed the less pretentious wooden affairs, and are erecting these mammoth skeletons of steel, securing greater strength and longevity and minimizing the possibilities of fire being communicated underground in case of the burning of the surface works. They are so built as to support the greatest weight at the least expense to the permanent stability of the structure and range in height from 110 to 125 feet.

Nicely poised at the top are two immense wheels, independent of each other, over which a woven belt or cable passes, connecting the cage with the hoisting apparatus. This belt is a powerfully made arrangement about one inch in thickness and about eight in width. So perfect is the action of this belt over the winding drum of the hoisting machine and the wheel of the gallows frame, and so sure the progress of the cage, either up or down, that the latter seems rather to be dropped to the bottom or shot to the top than to be handled like an elevator in our modern business buildings.

The novice on his first trip can thank his patron saint if his heart still beats at the completion of the same, providing the same degree of speed is given his particular cage as is given those handling the hardened miners. It will be observed that some of the accompanying illustrations of scenes below ground are not so good as others. The photographer responsible for the same was a novice in descending by the cage route, and some of the illustrations are products of his first trip.

In raising and lowering the cages to and from the different levels, broad stripes, painted on the outer side of the belt described, indicate to the engineer, sitting at his most responsible post, the exact location of either cage and permit the precise stoppage of the same at whatever level is desired. It might be mentioned that safety locks are placed at the top of each cage, which automatically fall out, in the event of the breaking of the belt or cable, holding the same and arresting its further descent.

Change is also noted elsewhere. In addition to the increase in the number of shafts or compartments thereto, increase also is made in the number

AFTER THE BLAST

Showing a section of drift or tunnel through solid granite, with no timbers.

IN THE STOPE.

Showing how timbers are used in constructing floors from level to level, while tearing out ore.

of cages to each shaft. Palpably two cages can not ascend or descend in opposite directions in the same compartment. To obviate further increase of compartments, "two decker" and "four decker" cages, or what is the same as two or four single cages fastened one on top of the other, are employed in many of the mines. In still other mines but two decks are thus utilized for cages, the space devoted in the four decker to the two remaining cages being used for a "skip." This skip is an arrangement not much different from what two cages would be if placed one on top of the other with the floor of the upper and the roof of the lower one removed and a sheet-iron or steel jacket placed on the outside of the whole. Thus, both skip or cages or both may be used for hoisting ore or waste to the surface while the cages may be used in raising and lowering miners.

In filling the skips at the different levels, a chute is dug a few feet back from each station in the center of the drift, descending for some feet and finding an outlet in the side of the shaft. Into this chute the ore is dumped and is released into the skip at the will of the attendant. In this manner both skip and cages are filled simultaneously. Once filled, signals are rung by an electric bell system into the engine-room and the skip and cages are elevated.

Arrived at the top, the cars are run off onto the tracks at the surface, after which the skip is hoisted to about twice its length above the surface, and, by an automatic arrangement, turns completely over and outward from the gallows frame, and empties its contents into a temporary bin immediately alongside of the structure. This system contemplates the use of larger cars, which, propelled by a steam locomotive of diminutive size, are backed under the temporary bin and transfer its contents to the larger ore bin. So perfect is the discipline maintained in the loading and unloading of cages and skips, and so well timed are these operations that almost to the second the signals to lower the one cage and raise the other are sounded in the engine room.

STATION

1100-foot level, Original Mine, three-compartment shaft in rear.

With the mine's development and the creation of immense tunnels wholly denuded of ore and abandoned, the waste matter, which carries no ore, is thrown into these deserted workings and the necessity of its elevation to the surface is obviated. Long before this period is reached, however, the waste dump has grown to immense

WORTHINGTON PUMP IN OPERATION, 1,100 FEET BELOW SURFACE.

AIR-DRILL.

Drilling into solid granite, 1,700-foot level, Anaconda Mine.

proportions, towering mountain high above the ground. The plan usually pursued is to erect an immense trestle, running off from the lower slope of the hill, along which a track is laid and gradually the whole trestle loses its shape, buried beneath tons and tons of waste matter dug from the bowels of the earth, of no possible value, yet silent witnesses to one of the world's greatest industries. Extension of these dumps are made from time to time, or entirely new ones are erected, running at a different angle.

With development have come other changes in the surface workings. New machinery of powerful capacity, and for every necessity has succeeded the old; with the increased demand for timber, large sawmills are erected at convenient access to the shafthouse or gallows frame, and millions of feet of lumber are cut into exact sizes and sent below to reinforce the battered walls.

Where once a series of stacks carried away the smoke from the mammoth furnaces, the single stack, from 100 to 125 feet in height, is gradually superseding them. And thus the development goes on. One mine, older or richer than the other, setting a new example and the remainder falling into line.

Below the surface, also, expansion and growth follow rapidly. Each day sees the drifts and crosscuts extended, the stoping pushed further to either side or higher up and the sump sunk to a deeper level. So consistently, so perseveringly is the system pursued that oftentimes a new shaft is sunk at the farthest opposite boun-

STEEL GALLOWS FRAME.
125 feet high.

STEAM ORE CARS.
Used in connection with steel gallows frame.

dary of the claim, and thus the work of digging, tearing, blowing out is prosecuted from both ends of the vein and the output thereby largely increased. In such cases or in groups of mines operated by one company the machinery of one mine is made to do duty for all, supplying fresh and compressed air, electricity, etc., to the levels of all.

One remarkable fact not common to all mining sections is that one can pass from one mine to another on the different levels for great distances. It is a truth that one can descend a shaft of a mine in Walkerville and ascend through the shaft of another at Meaderville, two miles or more away, without coming to the surface. So convenient is this system, due to the establishment of regular levels at given depths, that many surface workings of large mines have been wholly abandoned, even the ore being run into the levels of one mine

"FOUR-DECKER."

GOING DOWN.

centrally located and all hoisted through the one shaft.

In the case of the recent fire which destroyed the surface buildings of the Parrot mine, the miners, shut off from the raging flames at the very mouth of the shaft, found easy escape through the levels and shafts of no less than half a dozen different mines.

It is upon this fact of proximity and continuity of veins that so many mining suits of such tremendous proportions have been based, and which has made the term " apex " so common a word in the Butte vernacular. The generally recognized mining laws hold that the establishment of the fact that any given vein " apexes " in any certain claim gives the owner of that claim the right to work the whole of said vein wherever it takes him, if across the side boundary lines of such claim, although estopping him from proceeding beyond the end lines. With hundreds of claims, if not thousands, paralleling each other, some line of one serving as some line of another, the opportunity for irreconcilable differences in many instances at once suggests itself.

And thus in a general way proceeds the never-ceasing search and production of Butte's hidden treasures. By day and by night the work goes on — once the elusive vein is found — one shift following the other and taking up the work where it was left off. Each mine has its superintendent or foreman, and also its shift boss, whose duties include a continuous inspection of the work being performed throughout the mine, along the drifts and crosscuts, up in the stope and down in the low levels of the sump.

What the great body of men employed, working year in and year out, have accomplished for these many years the most active imagination fails to grasp. What a honeycomb of tunnels and shafts shooting in every conceivable direction lies beneath the surface of the small area surrounding Butte, braced and supported by millions and millions of feet of stanch timbers, no pen can describe, no picture show.

And yet a start only has been made. New machinery of greater power is being added to all the plants. Hoisting apparatus capable of raising or lowering cages from or to a depth of 4,000 or 5,000 feet are being installed, a depth not yet half attained, the average depth of the larger mines being about 2,000 feet, with a few reaching to the 2,300 level. Page upon page could be written of

AN INCLINE SHAFT.

specific incidents which would but reinforce the truth that the half has never yet been told concerning the possibilities of the future and increase the wonderment as to what the whole will reveal.

many operations entering into the production of copper do not end here. In fact they have but begun. It has been seen how, in a general way, the ore is extracted beneath the ground, elevated to the surface and finds its way to the dump or to the ore bin. The ore bin, in the mine's greater development, is not unlike a large grain elevator. On the side opposite from where the ore is emptied into it and some ten or twelve feet from the ground, large chutes, operated by cranks and gears, are raised and lowered and through these the ore is removed.

In most cases, standard gauge tracks have been run beneath the chutes, which enable ore cars, similar to the ordinary flat car, with sides and special unloading devices, to be run alongside, propelled by the ordinary switch engine. In other cases much smaller cars, propelled by electric power are used, and, in rare cases, either ore wagons or miniature ore cars, running upon narrow tracks, are utilized, drawn, respectively, by two and three teams or by a single horse. In cases where standard tracks are used, as the ore accumulates the cars are run in and loaded, and as rapidly as complete trains are made up they are hauled by powerful locomotives to the smelters controlled by the respective mining companies.

So complete has this system of railroad development proceeded that "the hill" is a perfect network of lines, running in every conceivable direction and at all manner of grades, the hill on closer inspection appearing to be terraced at every few yards by recurring tracks. Much smaller quantities of ore are handled by electric cars and never more than four cars constitute a train of this character. The amount of ore handled by horse-power is infinitesimal, and as the mine develops sufficiently to justify it steam power is substituted.

In addition to the numerous smelters operated in Butte, immense plants have been erected in Great Falls and Anaconda, and still further additions are in course of construction in the latter city, making that city easily the largest smelter town in the country, if not in the world. Trains consisting of from twenty to forty cars, carrying ore exclusively, are constantly following each other to the smelters of these cities or those in Butte, in the former case over the lines of railroad operated by the respective railroad companies.

ELECTRIC ORE CARS.
Operated through center of city.

Arrived at the smelter, the ore is again placed

STEAM ORE CARS.

in receiving bins. There are two kinds of ores, in point of quality — first and second class. First-class ore, according to the Butte classification, runs not less than seven per cent copper and is known as smelting ore or ore that is immediately melted without preliminary treatment. Second-class ore runs from two and one-half to seven per cent copper and is known as "concentrating" ore, and is sent to the concentrator. The purpose of concentration, plainly, is to eliminate a portion of the foreign matter and thus minimize the burden of the smelting department. As the treatment to which this class of ore is subjected precedes the smelting process, this phase will be considered first.

The principle employed throughout the concentration stages in every case is specific gravity, the specific gravity of mineral over the other ingredients being utilized to disintegrate the one from the other. The ore is first released from the bin through a chute and fed into the jaws of a powerful crusher, which reduces the rock

SIX-HORSE ORE TEAMS.

HORSE ORE CAR

to the approximate size of a walnut. The ore in turn then passes through succeeding sets of crushers, each reducing the size of the rock until it passes finally between two wheel crushers, the wheels revolving in opposite directions, which reduce the rock to about the size of sifted gravel. The ore is now run into jigs, at which stage the principle of specific gravity first is utilized. Water has been combined with the crushed rock and all is hydraulically forced through the troughs of the jigs, the silica being sufficiently light to be carried off, while sieves underneath the troughs allow a portion of the minerals, by reason of their specific gravity, to pass through, the jigs being given a motion similar to that which their name indicates to aid this operation.

The mineral thus abstracted is called "concentrates," and is conducted directly to the roasting furnaces, all necessity for further concentration palpably being obviated. While some mineral has thus been abstracted and some silica has been eliminated from the crushed rock, quantities of either still remain in the great bulk that has passed over the initial jig, and must be further concentrated. That which remains yet to be concentrated is called "middlings," and is carried on to the next series of jigs, there to undergo exactly the same treatment as in the initial one, resulting in the abstraction of some mineral and the elimination of some silica. And thus on, from one series of jigs to another, one a little lower than the other, the middlings are carried from floor to floor, each series performing its proportion of work. Finally is reached the Huntingdon crusher at the lowermost end of the jigs and into this the middlings are run

INITIAL CRUSHER.

HUNTINGDON CRUSHER.

and ground into a fine powder, not much coarser than flour.

Emerging from the Huntingdon the ore seems to have disappeared and muddy water to have been substituted. This is now conveyed to the "tables," which, likewise, utilize the principle of separation by specific gravity. The "round table" is the first to which the muddy water is run. It is an immense circular affair, its surface

CONCENTRATING JIGS.
Showing four floors devoted to these machines.

ROUND TABLES.
Water clearing on the right.

sloping uniformly from the center to the outer edge, and is given a revolving motion which never ceases. Large pipes run separately from above the center of the table, carrying, respectively, the muddy water and clear water.

The muddy water is released from the center upon a given half of the table under a nicely adjusted pressure and the clear water likewise upon the other half. As the muddy water passes downward over the surface of the table, the mineral, in its powdered condition, naturally flows less rapidly by reason of its greater weight and cleaves to the surface, while the water and less weighty ground substances flow with sufficient rapidity to pass completely from off the table and into receiving sluices provided therefor. Meanwhile the table, constantly turning, has carried the mineral remaining upon the surface around to the opposite side and over this the clear water is allowed to flow, eliminating still further portions of foreign matter not carried away by separation upon the initial half of the table.

As the table still further revolves, carrying the mineral with it, and just before it reaches the point where the muddy water will be poured upon it, a series of small waterspouts arranged above the surface of the table from center to outer edge, shoot strong streams across the surface, clearing it of the mineral as the table passes under. Thus a clean surface is constantly passing under the pipes carrying the muddy water and, automatically and without ceasing, the table is continually carrying its treasure of mineral around to the spouts to be swept off and treated as other concentrates.

Still another process is necessary, however, before the middlings are deprived of sufficient values to warrant a termination of further treatment by concentration. The middlings that have passed over the round table are next conducted through sluices to a different type of table, known as the Wilfley, in which the principle of specific gravity, differently applied, is used. These tables are long and narrow, with the foot a trifle lower than the head, and with a slight slant from side to side. A quick, jerking movement from end to end, similar to the jigs, is given these tables.

WILFLEY TABLE.
Showing separation of mineral and waste.

Along the surface from end to end and about

END OF CALCINE OVEN.
Showing plows emerging from inside.

HOPPERS.
Dumping roasted calcines into smelting furnace.

DRAWING OFF SLAG.

PREPARING TO FILL CONVERTER WITH CHARGE OF MATTE.

an inch apart are fastened most delicate strips of metal of barely perceptible thickness. As the middlings or muddy water is run from the round table it is brought to the Wilfley table and precipitated from along the side, near the head.

Here again the specific gravity of the mineral permits it to cling to the surface of the table, aided by the strips, while the foreign matter passes on over the table and is carried away to the final tailings dumps. Eight-tenths to one and two-tenths per cent of mineral is carried in these tailings and no machinery would pay for its operation in abstracting values from them. The mineral meanwhile adhering to the surface of the table gradually works its way to the foot and is held in its place by the strips; while constantly running water still further eliminates the foreign matter not washed away by the first separation of the tailings. As it reaches the end of the table, it passes over and, like the previously secured concentrates, is taken to the furnaces.

The average percentage of copper in all the concentrates secured from the concentrating process by means of the jigs and tables is about twelve per cent. It will thus be seen how by this process the operations following concentration are relieved of a vast amount of work. If the second-class ore, carrying from two and one-half to seven per cent, were taken directly to the smelter, as is the first-class ore, all of the foreign matter disposed of by concentration would have to be handled through the succeeding stages. Whereas, by eliminating it, the matter abstracted and taken to the smelter carries fully two to five times as large a percentage of copper. In other words, tons of ore turned into the crusher are taken away to the tailings dump and the necessity of smelting this great amount is obviated.

While much foreign matter has been separated from the ore, however, there still remains a great deal.

CONVERTERS BLOWING.
Showing immense electric crane on tracks at top. Furnaces to left of picture.

The foreign matter in Butte ores consists of from forty-five to seventy per cent silica and the balance of iron and sulphur.

The concentrates still contain but nine to fourteen per cent of copper, the remainder running about twenty to twenty-five per cent silica, forty per cent sulphur and the balance in iron and other base metals.

The next step is the elimination of the greater portion of the sulphur by roasting, which step is the first one encountered in the smelter proper. The concentrates are run through hoppers upon the beds of huge enclosed ovens, under which fires are constantly kept burning, the concentrates lying about two inches deep. These ovens are of a great length and lie usually one above the other. As the concentrates drop from the hopper into the oven, a plow-like device, with teeth a few inches apart, guided by wheels running upon tracks at either side of the oven and propelled by endless metal belts, scatter them along the surface and push them a little farther into the oven. At given intervals, other plows appear, turning the concentrates, pushing them a little further along and, passing on through the oven, follow the course of the belt to the oven below, performing a like service there.

Thus the plows continue their endless journey from one oven to the other, the concentrates gradually being deprived of the greater portion of the sulphur therein contained by virtue of the inflammable properties of that ingredient and the absence of such properties in other minerals. By the time the concentrates have been pushed through one oven, have dropped into the lower one and covered the length of that, the sulphur contained in them has been reduced to about eight or nine per cent. This being a sufficiently low per cent they are automatically pushed into waiting cars at the ends of the oven, and are carried to the " reverberatory furnaces."

The roasted ore is now known as " calcines." The reverberatory furnace is so called for the reason that the flames therein are made to reverberate and whirl. The calcines are dumped into hoppers directly above the furnaces and, as a slide in the hopper is pulled, the " charge " is dropped directly into the flames. The object of this process is obviously to melt or smelt the charge, the time taken in so doing ranging from four to six hours, according to its size.

THE FINAL STAGE.

Drawing off 99 per cent copper into pigs.

When the charge is thoroughly smelted the mineral, by reason of its specific gravity, seeks the bottom, while the waste matter, composed mostly now of silica and iron, rises to the top. This waste matter is called " slag," and is skimmed off into immense pots through holes in the front of the furnace. The slag pots, when filled, are lifted by immense electric cranes at the top of the building, deposited on electric cars and run to the waste dump. The matter still remaining in the furnace is called "matte." This matte is drawn off into cast-iron molds or tapped directly into the converters.

The converter is a huge iron pot, composed of two half shells. These shells, before using, are first lined with a deep bed of clay, some thirty inches in thickness, and are then fastened together. The crane then carries the converter to a point adjacent to the reverberatory furnace and is lowered into a hole sufficiently deep to allow the molten matter to run into it from the furnaces. The converter is now returned to its proper place beneath an exaggerated funnel-shaped pipe and compressed air is forced into the matte. This process is called "blowing."

As the blowing proceeds, the iron combines with the clay lining, forming a slag, which is poured off and the blowing continued; the sulphur, combining with the air, causes oxidation, and, presto, all foreign matter has disappeared and ninety-nine per cent pure copper remains. This is run off into molds, the bars being called copper pigs. These pigs are now too fine in copper to permit of treatment in local works and are shipped East to the refineries.

As "first-class" ores proceed directly from the mine to the smelter, the process to which they are subjected begins with the reverberatory furnace and their subsequent treatment is identically the same as the calcines from the second-class ores.

PUBLISHER'S NOTICE.

In submitting this humble effort, kindly thanks are publicly due to many who have assisted so generously in its production.

To August Christian, chief engineer of the Anaconda properties; John O'Neil, superintendent of the Anaconda, Neversweat and St. Lawrence mines; W. C. Thomas, superintendent of the Butte and Boston smelter, and Thomas Bryant, superintendent of the Original mine, special acknowledgments are due for having assisted in securing the most complete collection of mining and smelting views that, undoubtedly, has ever been assembled under one cover.

The engravings were made by the Illinois Engraving Company, of Chicago, whom we believe to be the peers in their line anywhere.

The paper was furnished by the Dwight Paper Company, of Chicago, and its high quality speaks for itself.

The composition, presswork and binding were performed by The Henry O. Shepard Company, of Chicago, printers of *The Inland Printer,* who require no eulogy from us.

Mr. Samuel Hamilton, of the Elite Studio, in Butte, has reflected his unquestioned ability as an artist in the character of the photographs, all of which are his handiwork.

No effort has been spared to provide a publication which every Montanian will be proud to see go beyond the State, and yet keep its selling price within so nominal a range as to make it a popular one and within the reach of all.

This publication may be secured from newsdealers and booksellers for $1.50 per copy, or will be sent by mail to any address in the United States or Canada, carefully wrapped, for $1.75.

HARRY C. FREEMAN,
Manager Montana Art View Company,
BUTTE, MONTANA, U. S. A.

HENNESSY MERCANTILE COMPANY.

Hennessy's, the "Biggest, Best and Busiest Store in Montana," is located on the southeast corner of Main and Granite streets, Butte. It is a brick building with steel framework and stone facings. It is an imposing structure, six stories high. The three upper floors, with the exception of a few rooms occupied by Hennessy's as store-rooms and offices, are rented by many of Butte's leading lawyers, physicians and professional men. The halls of these floors are covered with inlaid marble tiling. Fireproof vaults, for the use of tenants, are built in the solid masonry and occupy the center of each floor.

HENNESSY BLOCK.

The building, in its entirety, was put up in the most substantial manner possible, and is as near fireproof as human skill could construct, costing over $600,000. It measures 84 by 192 feet, is thoroughly lighted by electricity, and furnished with all the modern improvements.

Hennessy's store, about which so much has been said and written, occupies the three lower floors and the large, well-lighted basement extending under the sidewalks of Main and Granite streets. This store was first opened to the public on November 21, 1898, but the formal opening was deferred until Wednesday, December 7, some two weeks later, and was recognized as the most important mercantile event in the history of the State, marking the transition of Butte from a so-called mining camp to a metropolitan city.

It engendered confidence in the minds of Butte's citizens, who are now rapidly improving the city by the erection of handsome and substantial buildings, in both the business and residence portions.

Hennessy's store is the chief attraction in the city for shoppers from all parts of the State, who can save both time and money by the facilities furnished for supplying under one roof everything that everybody can need for their homes or personal use.

Heavy French plate-glass windows, framed in copper, extend the entire length and width of the building on both the first and second floors, furnishing admirable light to the interiors, and giving an opportunity for displaying goods that no other store in the State possesses. These windows are a sight in themselves. The main entrance to the store is on Main street. The grocery department at the rear is entered from Granite street, and between this and the main portion of the store is a handsome hallway entrance for the offices above, reached by electric elevator and Tennessee marble staircase with solid bronze balustrades, etc. The main floor of Hennessy's contains the silks and dress goods, domestics, notions, trimmings, hosiery and gloves, boots and shoes, men's clothing, men's hats, men's jewelry and furnishing departments, all of which are as complete in every detail as was possible to make them, and filled with stocks of goods that would make a good showing in any first-class store in New York or other large city, and which surpass any that can be seen in the Northwest.

SILKS AND DRESS GOODS DEPARTMENTS, ON MAIN FLOOR.

Step to the left and thousands of dollars' worth of silks, fresh from the looms of France, Switzerland and domestic points, fill the shelves, cover the counters and lend their graceful drapings to make a display of rich fabrics that can not be matched in the West.

Are you wanting an evening gown? There is a dark-room handy into which you can step to test the effect of electric light upon tints lovely by day and more or less so at night. Everything new in silks and velvets, imported trimmings and hand-made laces can be seen for the asking, and readily transformed in the dressmaking department to the richest reception gown that a reigning society belle could desire.

Pass on to the dress goods. Do you want a French novelty? It's here in a hundred styles.

Pneumatic tubes of polished brass connect the meat market, grocery and the several departments on the second and third floors and in the basement with the cashier's desk and wrapping departments in the center of the main floor.

Two passenger elevators and five freight elevators run by electricity are taxed to their utmost capacity, for, come when you may, you will find this store crowded and its three hundred and more employes busy attending to the wants of the many customers.

We said the departments were complete in every detail. They are, and noticeably so. As you enter from Main street you can not help

NOTIONS AND SHOES, ON MAIN FLOOR.

admiring the rich and highly polished quarter-sawed oak counters, fixtures and tables, and the show-cases of finest plate glass, showing off to the best advantage silks, laces, shoes, shirts or what not.

Do you admire the fashionable fabrics of finest wool or wool and silk for a tailormade suit? Here are the latest in Scotch, Irish and English tweeds, French venetians, broadcloths, zibelines,

MEN'S CLOTHING AND FURNISHINGS DEPARTMENT, ON MAIN FLOOR.

homespuns, golf suitings, covert cloths, camels'-hair serges, English diagonals, worsteds, and many other stylish materials for that comfortable costume. Many a swell tailormade garment has been correctly fashioned in Hennessy's dress-making department this season. Expert men-tailors from New York do the work and it's the finest.

LADIES' SUIT AND WRAP DEPARTMENT, ON SECOND FLOOR.

Are you a housekeeper? Then domestics will have some attraction, for pretty Irish linen table sets, Barnaby linen towelings, sheetings, bedspreads, muslins, English, French, Scotch, California and other flannels are shown in the greatest variety. Notions, the little things, but immensely important. The "Reynier" kid gloves, prizewinners at Paris this season, the "P. & L." and others in kid, silk, wool and lisle; hosiery, plain and fancy, laces, such a lot and so many styles. Isabel Cassidy's toilet preparations, soaps, perfumery, ribbons, belts and a million articles of everyday use are here.

FURNITURE DEPARTMENT, ON THIRD FLOOR.

Boots and shoes — Banister's — the best shoes made for men, shown in sixteen new styles, and Hennessy's celebrated EEEE (for ease). Shoes for women have made this department intensely popular. There's a style about the fit and finish of Hennessy's shoes that's hard to duplicate, and it shows up in the heavy walking as well as the lightest shoes for dress. Here are boys', misses' and children's shoes, rubber goods, mining boots and shoes and the celebrated "Workingman's Friend" and "Never Sweat" shoes, so well and favorably known throughout Montana.

THE MOORISH ROOM, ON THIRD FLOOR.

THE ART ROOM, IN BASEMENT BAZAAR.

The south side of the main floor is devoted to men's goods, the furnishings department, with its long row of plate-glass showcases, filled with shirts, neckwear, jewelry, etc., is particularly attractive. Examine the men's clothing. Such an assortment of rich styles in suits and overcoats, such a stock ot serviceable garments for business and workingmen you will not find elsewhere. The world's leading makes in underwear, hosiery, gloves, shirts and neckwear have a showing here that can not be duplicated, for here is done the biggest business in Butte. Knox hats and other well-known makes are shown in the latest shapes and tints.

The second floor contains everything for women and children and more too. Ready-to-wear costumes, tailor suits, dress skirts, tailor-made skirts, golf suits, silk petticoats, dress and shirt waists, corsets, underwear, boys' clothing and furnishings, jackets, capes, coats, furs, all are here, as also are the millinery, dressmaking, stationery and men's tailoring departments. Each is the best that money can buy or skill produce.

Then the third floor, covering a space of over 16,000 square feet, devoted to furniture, carpets and draperies. On this floor are sample pieces of furniture, hundreds of chairs, tables and other things, and no two are alike.

Immense warehouses down town hold the stock of furniture of which these pieces are but the samples. Hennessy's is the largest, richest, handsomest and best stock of furniture in Montana, and the biggest business in that line is done at Hennessy's.

When you have looked through the lines of carpets you will have noticed the richest Axminsters, Wiltons, moquettes and body brussels, and the lower-priced tapestries, ingrains and mattings. Here are lovely rugs from the mills of Pennsylvania and New York, and gems of oriental beauty from the hand looms of antiquity, shown in the Moorish room, with teak-wood stands, battle-axes, cushions and other requisites for cozy corners and home comforts generally. Hennessy's are house furnishers in the truest sense of the word. Compare the work on draperies, the fit of carpets or anything else with what has been done by others, and every time you will find Hennessy's the best.

Don't miss that art room in the basement bazaar! It's a gem, full of gems in cut-glass, fancy china, pretty bric-à-brac, graceful statuary, lovely lamps, pedestals, vases, tea and coffee sets, chafing-dishes, five o'clock teas and odd pieces of everything from everywhere. It seems to be the ideal spot in which to select a Christmas present, New Year's gift, something to beautify your own home or that of a bride about to establish one. Arranged on tables and shelves that meet the eye as you enter this section are china and crockery, glassware, stoves, heaters, toilet sets, dinner sets, hardware and the thousand and one big things and little things in house furnishings that every woman wants.

CROCKERY DEPARTMENT IN BASEMENT BAZAAR.

Then if something really good to eat and drink is wanted, there's no place like Hennessy's grocery for supplying that want. Most of Butte people appreciate this, if one can judge by the number of teams which deliver the goods to all parts of the city.

If you are a resident of Butte you can appreciate the advantages of dealing at Hennessy's, where goods are marked in plain figures, and twenty stores with the largest and best stocks in their line, valued at over half a million dollars, are under one roof, and so displayed as to make selections an easy task.

If you live in another part of Montana send in your order by mail. It will be filled promptly and carefully, and you will be given the advantages of low prices, of new and dependable goods at the Biggest, Best and Busiest Store in Butte.

KENNEDY FURNITURE COMPANY.

The Kennedy Furniture Company is one of the large commercial establishments of the city. It occupies the entire four floors of its retail store on West Broadway, and in addition has an immense warehouse. Both of these buildings are crowded with the most complete line of furniture to be found, probably, in the Northwest.

Not only do the lines of goods carried contemplate the more necessary articles of furniture ordinarily carried by smaller dealers, but extend into every branch of house-furnishing. So complete are these numberless departments that any person in any walk of life, from the wage-earner to the merchant, the bank-owner and the millionaire, can supply every want in the furnishing of his home, his office or store with the most minute article which could possibly suggest itself to his mind.

The Kennedy Furniture Company was organized on the ruins of the old Northwestern Furniture Company, in 1894. To the fine line of goods carried by this firm, consisting of stoves, ranges, crockery, bedding, etc., was added everything that possibly could be desired, until in a short time it occupied the position which it now enjoys — that of the leading furniture house in the Northwest.

One of the leading features of the stock carried by the Kennedy Furniture Company is its enormous line of carpets, rugs and tapestries, which equals any display that can be found anywhere, not excepting the large establishments of the East.

The motto of the firm, adopted at its birth, "The best goods, at the lowest possible prices, with fair, courteous treatment to all," has been so religiously observed by this concern that it

KENNEDY FURNITURE COMPANY

CARPET, RUG AND TAPESTRY DEPARTMENT.

CHAIR DEPARTMENT.

has gained the widest of reputations throughout the State for fair treatment, its trade extending to the farthest confines of Montana.

Responsibility is assumed for the statement that no better satisfaction can be secured anywhere in the Northwest, or even by going to the extreme East, than can be secured at this institution, and the people of Montana are learning that they can safely entrust their orders by mail to the Kennedy Furniture Company and have them filled, in cases where more limited stocks of home concerns are inadequate, quite as satisfactorily as if permitted to deal personally with the ordinary dealer.

BUTTE BREWERY.

Elsewhere has been shown the advantages to be gained by coming to the base of production of raw materials, both of mining, agriculture and sheep and cattle raising, for which Montana is growing famous. While a plea has been made for a larger activity in this direction, it should not be assumed that a start has not been made. On the contrary, many manufacturing institutions have been quick to see the benefits presented and are today receiving a rich reward for their far-sightedness.

In no instance is this more true than of the Butte Brewery. While an old landmark of the city, its life as an expanding institution is of recent birth. Although the raw materials used are not Butte productions they are among the many produced throughout the State, and in the large success attained by this institution lies a pungent moral which others might well consider.

The brewery was originally established by Henry Muntzer, in the year 1885, on Wyoming street, between Granite and Quartz, within a block of the busiest portion of the city. For many years Muntzer's beer enjoyed the highest local reputation, but it remained for the advent of Mr. J. V. Collins, as the head of the institution, to take its name and the excellence of its product beyond the city's confines and into the furthermost portions of the State.

Realizing the great future of Butte and the grand opportunity presented for greatly increasing the output of the plant, Mr. Collins, about a year ago, purchased the plant outright and immediately set about enlarging its capacity with the idea of making it the largest and most representative brewery in the State. That he is meeting a full measure of his anticipations is eloquently evidenced by the constant increase of output, and with the changes continually being inaugurated the time is not far distant when his most sanguine ideas will have been realized. Experts claim that the product now equals, if it does not excel, any foreign article, and the logic of such a claim is not far to seek.

All materials used in the making of the Butte Brewery's beer are grown either in Montana or the Pacific States — the barley in the former and the hops in the latter — and the fact that the largest brewers of the United States, as well as those of Germany, are exerting every effort to secure the entire products of these sections as the best of their kind in the world, speaks with eloquent emphasis of their superior qualities.

A fact which but few beer drinkers know is that, before beer can be shipped for great distances and subjected to severe handling and constantly changing temperatures, it is necessary to fortify it by the use of unsanitary drugs, else it would be undrinkable at its destination. Equipped as the Butte Brewery is with the latest machinery and employing precisely the same methods as the large foreign brewers, it becomes immediately apparent why the above referred to claims are unanswerably true, it being palpable that the necessity for the use of drugs in this brewery's product is entirely eliminated. By virtue of its proximity to its field of distribution, it is possible to draw it from the cellars to the kegs and for it to be consumed in the same day, thus obviating its subjection to abnormal temperature changes. These facts alone should be sufficient to show the unprejudiced the fallacy of crying for Eastern beer.

Mr. Collins, the president of the company, is an old-time Montanian, having come to Butte in the spring of 1884. He was for many years the manager for H. L. Frank, and for the past six years has conducted a wholesale liquor store and the Pabst beer agency in Anaconda. By virtue of his extended experience, Mr. Collins is thoroughly equipped to handle the financial and business affairs of the concern, and in the technical details of brewing he is most ably advised and assisted by Mr. E. W. Walsh, a thorough gentleman and an experienced brewer, who serves as secretary-treasurer of the concern, and of whom the trade says he is second to none.

Immediately upon his assumption of the control of the plant, Mr. Collins set about him to reconstruct, tear down, build up and increase in every way the possibilities of the brewery, to the end that the highest quality of beer might be produced. His efforts were not long without reward, for from the very beginning of his control the output of the plant has steadily increased, and the fame of the product has worked its way over the entire State, until the institution was taxed to its utmost to meet the demand.

Anticipating some time ago the trend of the new conditions, Mr. Collins foresaw what this all meant and set about to meet the situation. Plans have been drawn up to entirely rehabilitate the old structure, which the next year will see completed, and in the place of the old will be found an apparently new institution, more modern, more scientifically constructed and with its capacity more than doubled.

First, there will be a new brewhouse, with a capacity of 125 barrels a day. This, of course, means an entirely new brew outfit throughout, the present one being wholly inadequate to meet even the present demands, to say nothing of the growing ones of the future. New cellars, naturally, must follow to meet these new conditions, and such additions contemplate a capacity twice as great as those at present in use. A malthouse, too, is on the list of improvements, and its new capacity will approximate about 10,000 pounds per day. In fact, an entirely new brewery will have taken the place of the old before the year has rolled around, as large, complete and up to date as any in the State, if not more so.

BUTTE BREWERY.

Malt room. Filling room. Mash tub.
Brew kettle. Cellar, showing chip casks.
Fermenting room. Bottling house.

A FEW DEPARTMENTS OF THE BUTTE BREWERY.

The watchword of the brewery management, as it has been since taking up the reins of control, will be the best brewery in the country and the completest plant, conducted upon honest, upright methods, and, with the combination of brains and business sagacity at its head, it is safe to say that when time has a little further elapsed, it will become manifest to all that the half has never yet been told in point of prophecy concerning its future.

THE GREATEST TRANS-CONTINENTAL RAILROAD.

As elsewhere shown, Butte is adequately supplied with railroads. There is one among them, however, easily occupying the place of prestige, and not only Butte, but all Montana, is in accord in yielding that prestige to the Northern Pacific Railway. While neither terminal of that system is found within the State, it nevertheless is the most distinctively Montana railroad entering the city of Butte or tapping the most resourceful sections of the State. Following closely the growth of the great Northwest, from the time its rails first connected the Great Lakes with the Pacific coast, today finds it one of the completest railway systems in the United States.

Starting from St. Paul, it taps the best agricultural sections of Minnesota and North Dakota, has an absolute monopoly by the time Montana is reached, by exclusively touching nearly every desirable portion of the State, and, on to Portland, Seattle and Tacoma, runs through the more thickly settled sections, giving the traveler a better idea of the resources of the country traversed than can be secured by any other route across the country.

Its equipment is new and bright; its roadbed, gradually straightened by constant labor, presents none of the sharp curves of a few years ago, and except where contour of river, valley or mountain makes it impossible, pursues an air-line course for miles and miles. Its rails, too, have increased in weight and the rolling stock glides over them so easily as to entirely obviate the nerve-racking, body-tiring jerks so common to many so-called railroads of the first class.

Science in railroad building has even gone further. Where once existed many trestles, now is found a roadbed and gracefully sloping bank identical with the remainder of the road. In all cases where the trestle was advantageously placed, immense hydraulic apparatus was set in operation on the higher side of a gorge and a large portion of the mountain washed away, the earth being carried by the force of the water to the deeper levels of the gorge. Here it settled into all the nooks and corners and, accumulating, gradually buried the trestle beneath it until the last beam had been covered, and a bed, as firm as any that nature had built, rested beneath the rails. In scores of instances new wooden trestles have first been erected only to be thus buried beneath their load of earth, the result being that no railroad in the country today boasts of a more secure roadbed, more free from danger of fire or washout, than is the Northern Pacific road.

Not only is its line the only trans-continental line entering directly into the city of Butte, but it is also the only one enjoying that distinction in nearly every one of the remaining prominent cities of Montana. Anaconda, the great smelter city; Helena, the capital; Missoula, in the heart of the noted Bitter Root valley; Bozeman, the metropolis of the Gallatin valley, famous for its wonderful barley; Livingston, the gateway to the matchless Yellowstone Park, and Billings and Glendive, from whence are shipped the cattle and sheep whose delicious qualities have given Montana an enviable reputation as a great stock country; all these, running midway across the State and adjacent to every great industry of Montana, know the Northern Pacific as their greatest railroad, in most cases enjoying the presence of no other — nor seeking it.

Recently there has been added to the equipment of the system the last feature necessary to destroy whatever distinction might have existed between the Northern Pacific and its great Eastern contemporaries — the Pennsylvania and the New York Central systems. This feature consisted of the latest innovation in railroad comfort — the Observation car — and made of the "North Coast Limited" the most magnificent trans-continental express running between the East and the West over any line. A more complete and detailed description of this great railway achievement should be included in a description of the train as a whole.

Something that will please the overland traveler as a happy improvement over previous conditions is the elimination of frequent stops at unimportant points. For hours at a time this wonderful train pounds across the country, over mountain and valley alike, past village after village, without so much as a slackening of speed, drawn by powerful engines, and the impression at the conclusion of a journey covering thousands of miles is that not over half a dozen stops could possibly have been made. Another improvement, heretofore commonly reserved for the millennium, is, by way of an aside, the entire and total absence of courteous treatment " for revenue only " upon the part of attendants, the writer for the first time in a dozen or more similar journeys feeling free to command the services of respective attachés at will without the spur of financial consideration. In not a single instance was service rendered in any spirit other than that the attendant was present to minister to the comforts of the traveler, and the feeling that the more desirable comforts were reserved for the few was in no instance experienced. Whatever the cause may be is not known, but the effect is a most pronounced and acceptable innovation. It, for the first time, loaned realism to the comforts manifestly intended by the originators of " palace cars," and without which one so often wishes he had remained at home.

NORTH COAST LIMITED.
Commodious observation platform at rear.

If possible, the sleeping car is an improvement. The seats seem more designed for comfort — a curve here for the ease of the arm or hand, a cushion there for cheek or head prolongs the time before bodily aches common to long journeys present themselves, and the journey, presto, is over before the ache is located. The smoothness and air line directness of the roadbed, too, lends its beneficial effects to architectural comforts and accounts largely, no doubt, for this result.

Electricity has forced its usefulness upon the comforts encountered elsewhere and, besides the brilliant rays shed from the many chandeliers running the length of the train, each berth is supplied with a bulb at either end. The necessity which heretofore compelled the traveler to lay aside his novel and disrobe amid impenetrable darkness at the bidding of the all-mighty porter has gone the way of other early inconveniences and the traveler for the first time experiences the delights of "reading himself to sleep" aboard train. No noise of hilarity nor the fumes of foul tobacco now find their way to the ear or nostril of the would-be sleeper to disturb his slumbers. The smoking-room, formerly used for cards, conversation and smoking at the far end of the sleeping car, could be entirely dispensed with now so far as its use for any of the purposes originally intended are concerned. This is due to the presence upon the same train of an innovation so much more commodious, comfortable and sanitary as to make the once indispensable smoking-room a thing to be avoided, if only by contrast.

This innovation is the previously mentioned observation car, which is attached to the rear end of the train, immediately next to the sleeping cars. This car is more, by far, than its name implies. It is a combination of everything that can lend bodily or mental comfort to the traveler. If he would play cards, two commodious rooms, each electric lighted, containing a half-dozen wicker chairs movable at will, which, in turn, surround a regulation card table, are provided for this exclusive purpose. A complete buffet, attended by a willing porter, adjoins these rooms, and solid or liquid refreshments are promptly to be had at any hour of the day and part of the night.

A corridor leading from the train end of this car passes these card rooms and, continuing, brings the traveler upon a scene and a suggestion of comfort rarely encountered upon a train in any portion of the country and the first to be seen on a Western road. Before him lies a complete barber shop, presided over by the best of artists in his line and equipped with great lockers of snowy linen and every instrument and convenience common to a first-class city shop. By a glimpse through the open door to the left of the shop one experiences a still further sensation and promise of comforts to come in the presence of the neatest of little bathrooms, equipped, as is the shop, with everything possible to make it complete — perfect seclusion, electric lights, hot and cold water, brushes, showers, the whitest of linen and perfect ventilation. Surely the question of comfort seems to have been exhausted.

Proceeding along the corridor, however, toward the rear, the traveler finds that others

PALACE SLEEPING CAR, ELECTRIC LIGHTED.

OBSERVATION CAR.

Showing portion of parlor, with stationary seats, library, desk, etc., at farther end.

TOURIST SLEEPING CAR.

GLIMPSES OF INTERIORS OF THE NORTH COAST LIMITED.

more thoughtful than himself have studied out this question. As the corridor takes a graceful turn, he sees before him the embodiment of sufficient aids to comfort to fill three ordinary cars. Coming as he does into the center of the car, he finds himself in the coziest of little libraries. Here is a bookcase filled with all the latest magazines bound in soft leather covers, together with every conceivable literary work adapted to train reading, capable of being picked up or laid down at the will of the reader and as diversion demands. A delightfully appointed desk is a part of the library, which fills a long-felt necessity, and here the traveler finds it possible to write in comfort with everything at his fingers' ends necessary to do so, with the ease and facility enjoyed in his home or office. A mail box, even, is provided, from which the mail is taken and attended to by the omnipresent porter.

Passing from the library, the traveler encounters stationary seats, upholstered in a rich material of soft green hue in harmony with the general coloring of the interior of the whole car. Beyond these stationary seats one emerges into a parlor as exquisitely appointed as those of the modern hotel. Soft Wilton rugs cover the floors, and large, inviting wicker chairs, of different sizes and shapes, upholstered in harmonious colors, are distributed the length of the room and are revelations of comfort. The windows — huge plate glass affairs some four feet square — are hung with shades to match and permit of an easy and advantageous indoor observation of the fleeing landscape not enjoyed from the ordinary car window.

But the greatest feature of all remains. Beyond the parlor and through the rear door of the car is the observation platform. For a space of about seven feet beyond the door and the width of the car the floor is extended, and from either side around the outer edge a high brass railing of artistic design is run. The platform floor is covered with some ornamental material resembling tiling or marble, an electric light is suspended from an arched dome above and folding camp chairs are numerously provided. Here, removed from smoke and dust, with the landscape running away from either side of the car, one sits for hours enjoying the delicious air, the sunshine, the ever changing scenes, quietly smoking his cigar or pipe, reading his novel or engaged in conversation, or, totally absorbed in the very charm presented, remaining perfectly silent and drinking this new cup of happiness to the full. Plainly the "sleeping" car has acquired a new significance in that, henceforth, it will be known and used as the "bed chamber" of a now complete train.

With the refreshing sleep to be obtained from the improved accommodations and the absence of objectionable features noted in the sleeping car, and the invigorated condition created by the diversified comforts of the observation car, comes the increased appetite. One hesitates in noting improvement in the dining cars, for time out of mind the fare spread before the traveler on this road has far surpassed the most extravagant expectations, and but little room for improvement existed. One feature most acceptable is the serving of breakfast and luncheon *a la carte,* enabling the traveler to satisfy his needs at an expense in proportion to them, orders costing as little as 25 cents being served in the same first-class manner as the highest-priced meals. Constantly revolving electric fans cool the car and keep it at a refreshing temperature, while fresh flowers adorn the various tables. Here, as elsewhere in the car, that remarkable something has been at work, eliminating the "courtesy for revenue only" feature of the service and one finds himself as carefully looked after as if attended by his personal manservant.

Another feature, although of long standing, is the tourist sleeping car. A potent deterrent to overland travel in the present day is more largely due to ignorance concerning the reasonableness with which such journeys can be made than, perhaps, any other one cause. For the purpose of carrying large families, homeseekers of modest means, etc., at a much more reasonable figure than travel by regular Pullman service entails, the Northern Pacific has introduced a rate to especially meet these conditions. For the benefit of persons so traveling, exact counterparts of the Pullman sleeping cars are included in the train, the only difference being that the former are finished in fine furniture leather rather than in softer draperies. A Pullman porter presides over the car exclusively and, to the slightest detail, every necessary comfort enjoyed in the more expensive Pullman is here present. The berths are prepared by the porter and the whitest of linen is furnished. Travelers by this car enjoy the privileges of the Pullman dining car, and the only difference between the two methods of travel is the reduced rail and sleeping-car fares.

A first-class day coach is also run upon this train as well as the regulation smoking, baggage and mail cars and, from end to end, the train is probably as complete, comfortable and delightful as any train running in the United States today, and by far and away is this true as concerns any train crossing the continent.

ELECTRIC LIGHTED DINING CARS.

What the home presents and not found here is a humane omission, fraught with worry and care and not conducive to a pleasure journey; while everything that lends comfort to the body, pleasure to the inner man, rest and quiet to the distracted mind and food to the very soul are here all present, and the traveler arrives at the journey's end, whether it be Seattle or St. Paul, rested and invigorated, hardly realizing that thousands of miles have been stretched between himself and his starting-point.

CHICAGO GREAT WESTERN.

THE LEADING EASTERN LINE.

Following is an illustrated description of the "Great Western Limited," the evening express from St. Paul to Chicago over the Chicago Great Western, which connects with the North Coast Limited at the former city. This train is run over the line of the Chicago Great Western Railway and has a peculiar interest to Montana readers in general and those of Butte in particular, in that this line has a general agency at 15 West Broadway, in Butte, and is therefore a part of Butte's institutions.

To the traveler whose journey's end lies beyond the eastern terminal of the Northern Pacific at St. Paul, is presented the new question of a route to his destination. With a mind quickened with the knowledge of the good things in railroad travel but recently enjoyed on the elegant North Coast Limited, and a lingering relish for their continued enjoyment, the question's solution is freighted with no little need for consideration. Five hours of waiting and an all night's ride to Chicago confront him in whatever direction he turns. What an acceptable five hours for recreation and a general stretching of limbs and muscles preparatory to the continuation of the journey, free from all thought concerning this momentous question, provided its solution is reached in advance. If the experiences of another, likewise situated and keenly anxious to thus employ these hours, aids in any way the solution of this question for the traveler, the purpose of this article will have been happily realized.

Arriving in St. Paul at three o'clock in the afternoon, the shadows of night will long have fallen before the time of departure eastward. The journey then resolves itself not into one of sightseeing, but rather presents the question of how best to spend the evening, to secure a refreshing sleep and obtain an enjoyable morning meal before leaving the train in Chicago. If this be the end sought as the happiest solution, then full responsibility is assumed for the statement that nowhere can this consummation be attained so completely as by continuing the journey over the evening train of the Chicago Great Western line.

Here are found two complete innovations — an elegantly appointed buffet car for gentlemen, and a perfect revelation of a compartment car — with a free chair car to supplement the day coach of the North Coast Limited. The compartment car is a novelty in palace-car construction. It could be likened to a living floor of a modern

Compartment Car Corner — Within the seclusion of four walls.

hotel. A corridor runs its entire length, flanked upon one side by a series of staterooms or compartments, each complete within itself and capa-

COMFORTABLE CHAIR CAR.

ble, like a hotel room, of being entirely shut off from all other parts of the car or thrown together en suite. Doors connect each stateroom, not only with the main corridor, but open upon either side into the adjoining compartment through heavily mirrored doors. These doors in all instances, however, may be locked at the will of the traveler, or a number of compartments may be thrown together for the accommodation of many members of a family or a large party traveling together.

The initiated traveler will find in these compartments the embodiment of every essential of the drawing-room of the regulation sleeper, with toilet, hot and cold water for bathing purposes, ice-cold drinking water, mirrors, ample floor

BUFFET CAR.
A Home Parlor.

space for convenient disrobing or dressing, independent of the berth, and the delightful privacy which the drawing-room permits. The traveler who profits by the well-intended suggestion here offered will have his confidence rewarded within the first five minutes' investigation of this most ideal departure in car building, and will have nothing but words of praise for the one responsible for the suggestion.

Safely ensconced in his quarters, privilege is afforded him to leisurely arrange his luggage to suit no one's convenience but his own, to remove the effects of dust and heat acquired by his recreation between trains by a dip in his private washbowl, to dress his hair and don fresh linen and a more comfortable coat and hat, and, if so moved, to light a good cigar without fear of restraint from his neighbors. For he is absolutely alone and is controlled by nothing save his own wishes and his own comforts. A four-burner gas chandelier is provided in each compartment and a pleasant hour or two may be devoted to uninterrupted reading with all the comfort of the

LUXURIOUS SLEEPING CAR.

traveler's own parlor if he so wills. If, perchance, the inner man presents sufficient argument to entitle him to consideration, the traveler may pass through the train and into the buffet car, where an obliging waiter will quickly furnish him with a most appetizing lunch of cold or potted meats, sandwiches and relishes of all kinds. Liquid refreshments of the sparkling kind or the more domestic cup of coffee may also be had for the asking, all of which may either be served at an individual table in the adjoining lounging and reading room or in his private compartment.

If accompanied by his wife he may even secure all of this by the slightest pressure of a bell at the side of his seat in the compartment car, whereat an accommodating porter will do the rest. If he elects to enjoy his lunch in the lounging room of the buffet car, long before its discussion is completed he has become so enamored of the home-like atmosphere everywhere surrounding him, that it is a safe guess that he will remain and read until too sleepy to do so longer.

Few metropolitan clubs furnish a more delightful corner than does this library-buffet car in which to while away the hours of an evening, smoking, reading or chatting.

For those who so desire, of course, the regulation Pullman sleeper is at their disposal on this train and is as complete and comfortable as those the traveler enjoyed on the North Coast Limited. On this line, also, the roadbed has been brought to the highest degree of perfection, and the sleep enjoyed by the traveler in consequence thereof is as refreshing as any strange bed permits, and with the coming of the morning he is profoundly conscious of an ability to eat the better portion of a Montana beef. If anything is needed to give zest to his appetite, it is supplied in the excellence of the fare itself. After its full discussion the traveler, again finding his way to the buffet car to enjoy his morning paper and after-breakfast smoke, is fully prepared to agree that no trip of a similar character was ever so pleasantly spent; and, as he alights at the magnificent Grand Central station in Chicago, the Chicago terminal of the Chicago Great Western, an hour later, conscious of having spent less than $4 for all the privileges enjoyed, he at least promises himself a repetition of the same in the near future.

Sitting-room, bedroom, toilet room and privacy in one.

INVITING DINING CAR.

With courteous attention.

Promising themselves a repetition of the journey.

The Worthington Pump

THE cut herewith shown gives an excellent idea of the above named pump. It is one of an endless line of like machines made for every purpose and of every size and capacity.

This pump is located at the 1,100 level of the original mine, Butte, Montana, and is pumping 200 gallons of water per minute against a vertical head of 1,050 feet. It is one of the features of deep mining. Too much attention can not be paid to having reliable machinery for this work, as a stoppage of the pumps would mean a loss of thousands of dollars as well as jeopardizing the safety of the mine and miners.

This type of an engine is accepted and used in Butte mines generally as representing the highest degree of economy and durability.

The water end is made entirely of phosphor-bronze, which makes the first cost of such an engine much larger than those of inferior qualities. The extreme long life of this pump, as well as all others manufactured by HENRY R. WORTHINGTON, which this superior material assures, more than compensates for first cost in a total absence of supplementary costs which are bound to attend an inferior grade. The saving in fuel alone over the regular compound condensing engine is very great, while the machine is so simple that it requires no more expensive labor in its care.

Pumps of this kind can be built for any service and contracts made with guarantee of duty.

The accompanying cut is significant in that it shows that the Worthington Pump is one that gets outside of the warehouse and into active use in the largest fields, where absolutely perfect results are required.

Carlisle Mason, at No. 110 North Wyoming Street, is the Butte Sales Manager for HENRY R. WORTHINGTON, the manufacturer of the Worthington Pump. The Home Office is in New York City, while Branch Offices are located at Boston, Philadelphia, Chicago, St. Louis, Cleveland, Detroit, Atlanta, Pittsburg and New Orleans.

MONTANA MUSIC CO

SALESROOM, MONTANA MUSIC CO.

DANDRUFF IS A GERM DISEASE

Dandruff

Means
Itching Scalp;
Thin, Brittle and
Falling Hair; and,
Finally, Baldness.

Without Dandruff,
There'd be
No Falling Hair,
No Baldness;
And Hair
Would Grow
Luxuriantly.

"Destroy the
Cause, You
Remove the
Effect."

Use
"Herpicide."

You'll Have
Dandruff
All Your Life,
Unless You
Kill the
Dandruff Germ.

You Can't
Do that Unless
You Use

Newbro's "Herpicide,"

The Only Hair
Preparation
That Actually
Does Kill the
Dandruff Germ.

Allays Itching.

Makes Hair
Soft as Silk.

Dandruff is a Germ Disease

"Herpicide," $1.00 At All Druggists

Kill the Dandruff Germ